Information and Instructio

This shop manual contains several sections each covering a specific group of wheel type tractors. The Tab Index on the preceding page can be used to locate the section pertaining to each group of tractors. Each section contains the necessary specifications and the brief but terse procedural data needed by a mechanic when repairing a tractor on which he has had no previous actual experience.

Within each section, the material is arranged in a systematic order beginning with an index which is followed immediately by a Table of Condensed Service Specifications. These specifications include dimensions, fits, clearances and timing instructions. Next in order of arrangement is the procedures paragraphs.

In the procedures paragraphs, the order of presentation starts with the front axle system and steering and proceeding toward the rear axle. The last paragraphs are devoted to the power take-off and power lift systems. Interspersed w
specifications pertai

HOW TO USE THE INDEX

Suppose you want to know the procedure for R&R (remove and reinstall) of the engine camshaft. Your first step is to look in the index under the main heading of ENGINE until you find the entry "Camshaft." Now read to the right where under the column covering the tractor you are repairing, you will find a number which indicates the beginning paragraph pertaining to the camshaft. To locate this wanted paragraph in the manual, turn the pages until the running index appearing on the top outside corner of each page contains the number you are seeking. In this paragraph you will find the information concerning the removal of the camshaft.

More information available at haynes.com
Phone: 805-498-6703

Haynes UK
Sparkford Nr Yeovil
Somerset BA22 7JJ England

Haynes North America, Inc
859 Lawrence Drive
Newbury Park
California 91320 USA

ISBN-10: 0-87288-092-3
ISBN-13: 978-0-87288-092-4

Disclaimer

There are risks associated with automotive repairs. The ability to make repairs depends on the individual's skill, experience and proper tools. Individuals should act with due care and acknowledge and assume the risk of performing automotive repairs.

The purpose of this manual is to provide comprehensive, useful and accessible automotive repair information, to help you get the best value from your vehicle. However, this manual is not a substitute for a professional certified technician or mechanic.

This repair manual is produced by a third party and is not associated with an individual vehicle manufacturer. If there is any doubt or discrepancy between this manual and the owner's manual or the factory service manual, please refer to the factory service manual or seek assistance from a professional certified technician or mechanic.

Even though we have prepared this manual with extreme care and every attempt is made to ensure that the information in this manual is correct, neither the publisher nor the author can accept responsibility for loss, damage or injury caused by any errors in, or omissions from, the information given.

SHOP MANUAL
FORD

SERIES

501-600-601-700-701-800-801-900-901-1801-2000-4000

The Tractor Model and Serial Numbers are stamped on the top of the transmission housing at the left front corner. A different model number is assigned each major product option. Early agricultural and industrial model number consists of three digits, followed in some cases by a suffix consisting of a number and/or letter. The first digit designates the engine size and tractor type, the second digit the transmission type and the third digit the year range of the series. The following table lists the product options indicated by the model number:

5**—One-row, 8" offset design equipped with 134 cu. in. gasoline or LP-Gas or 144 cu. in. diesel engine.

6**—Four-wheel, adjustable axle design equipped with 134 cu. in. gasoline or LP-Gas or 144 cu. in. diesel engine.

7**—High clearance row-crop type equipped with 134 cu. in. gasoline or LP-Gas or 144 cu. in. diesel engine.

8**—Four-wheel adjustable axle design equipped with 172 cu. in. engine.

9**—High clearance row-crop type equipped with 172 cu. in. engine.

18**—Four-wheel, axle type, industrial tractor equipped with 172 cu. in. engine.

1—"Select-O-Speed" transmission without pto.

2—Four-speed transmission without pto or hydraulic lift.

3—Four-speed transmission without pto.

4—Four-speed transmission.

5—Five-speed transmission with transmission pto

6—Five-speed transmission with live pto.

7—"Select-O-Speed" transmission with single speed pto.

8—"Select-O-Speed" transmission with two-speed and ground drive pto.

**0—Series designation built 1955 to 1958.

**1—Series designation built 1958 to 1962.

***-1—Tricycle type with single front wheel.

***-4—High clearance, four-wheel, adjustable axle type.

***-D—Diesel engine.

***-L—LP-Gas engine.

***-37—Equipped with Reversing transmission.

***-21—Equipped with Combination transmission.

Late agricultural and industrial model number consists of five digits, followed in some cases by a suffix consisting of a number and/or letter. The first digit designates engine size; the second digit, successive models; the third and fourth digits, tractor type; and the fifth digit, product options including transmission type. (NOTE: Options indicated by fifth digit varies with successive models.) Suffix letters and numbers are similar to those used for early agricultural and industrial types. The following table lists product options indicated by the model number:

2****—Indicates 134 cu. in. gasoline or 144 cu. in. diesel engine.

4****—Indicates 172 cu. in. gasoline or diesel engine.

*0***—Indicates industrial models produced prior to 1963.

*1***—Indicates agricultural and industrial models produced in 1963 and later.

**10*—High clearance agricultural Row Crop type, with single or dual tricycle front wheels or wide adjustable front axle.

**11*—Offset four-wheel agricultural type for one-row cultivation.

**20*—Four-wheel All-Purpose type with adjustable front axle.

**21*—Four-wheel orchard and grove type with non-adjustable front axle.

**30*—Four-wheel utility type industrial with non-adjustable front axle.

**31*—Four-wheel low center of gravity type with adjustable front axle.

**41*—Heavy duty industrial type with sub frame and cast grille, extra heavy front axle & steering.

*1**0—Four-speed transmission without pto.

*0**1—Four-speed transmission without pto or hydraulic system.

*1**1—Four-speed transmission with pto.

*0**2—Four-speed transmission and hydraulic system, without pto.

*1**2—Five-speed transmission with live pto.

*0**3—Four-speed transmission, hydraulic system and pto.

*0**4—Select-O-Speed transmission without hydraulic system or pto.

*1**4—Select-O-Speed transmission without pto.

*0**5—Select-O-Speed transmission, hydraulic system and 540 rpm pto.

*1**5—Select-O-Speed transmission with 540 rpm independent pto.

*0**6—Select-O-Speed transmission, hydraulic system and 540-1000 rpm and ground speed pto.

*1**6—Select-O-Speed transmission with 540 and 1000 rpm independent pto.

*1**7—Select-O-Speed transmission with 540 and 1000 rpm independent and ground speed pto.

INDEX (By Starting Paragraph)

INDEX (Continued)

CONDENSED SERVICE DATA

TRACTOR MODELS	600, 700	501, 601, 701, 2030, 2031, 2110, 2120, 2130, 2131	800, 900	801, 901, 1801, 4030, 4040, 4110, 4120, 4121, 4130, 4131	501D, 601D, 701D, 2030D, 2031D, 2110D, 2120D, 2130D, 2131D	801D, 901D, 1801D, 4030D, 4040D, 4110D, 4120D, 4121D, 4130D, 4131D
GENERAL						
Engine Make	Own	Own	Own	Own	Own	Own
Cylinders	4	4	4	4	4	4
Bore—Inches	3.44	3.44	3.90	3.90	3.56	3.90
Stroke—Inches	3.60	3.60	3.60	3.60	3.60	3.60
Displacement—Cubic Inches	134	134	172	172	144	172
Compression Ratio	6.6	7.5	6.75	(1)	16.0	16.0
Pistons Removed From:	Above	Above	Above	Above	Above	Above
Main & Rod Bearings Adjustable?	No	No	No	No	No	No
Generator & Starter Make	Ford					
Carburetor Make (Gasoline)	Marvel-Schebler TSX				....	....
Carburetor Make (LP-Gas)	Zenith				....	....
Distributor Make	Ford				....	....
TUNE-UP						
Compression, Gage Lbs.	120-130	130-140	120-130	130-140 (2)	365-400	365-400
Firing Order	1-2-4-3	1-2-4-3	1-2-4-3	1-2-4-3	1-2-4-3	1-2-4-3
Valve Tappet Gap—Intake & Exhaust	0.015	0.015	0.015	0.015	0.015	0.015
Valve Face Angle—Degrees	43½	43½	43½	43½	43½	43½
Valve Seat Angle—Degrees	45	45	45	45	45	45
Ignition Timing	See Paragraph 159				....	....
Injection Timing	....	....	....	....	See Paragraph 139	
Spark Plug Make	Autolite				....	....
Spark Plug Model (Gasoline)	AL7T	AL7T	AL7T	AL7T	....	....
(LP-Gas)	ATL3A	ATL3A	ATL3A	ATL3A	....	....
Engine Low Idle—RPM	475	475	475	475	675	675
Engine High Idle—RPM (4-Speed)	2250	2250	2250	2250	2250	2250
(5-Speed or "Select-O-Speed")	2450	2450	2450	2450	2450	2450
Battery Terminal Grounded	Positive	Positive	Positive	Positive	Negative	Negative
SIZES—CAPACITIES—CLEARANCES						
Crankshaft Journal Diameter	2.4981	2.4981	2.4981	2.4981	2.4973	2.4973
Crankpin Diameter	2.2985	2.2985	2.2985	2.2985	2.2985	2.2985
Camshaft Journals Diameter	1.9255	1.9255	1.9255	1.9255	1.9255	1.9255
Piston Pin Diameter	0.9122	0.9122	0.9122	0.9122	1.1242	1.2492
Valve Stem Diameter, Intake	0.342	0.342	0.342	0.342	0.342	0.342
Valve Stem Diameter, Exhaust	0.341	0.341	0.341	0.341	0.341	0.341
Main Bearings Running Clearance	Refer to Paragraph 109					
Rod Bearings Running Clearance	Refer to Paragraph 108					
Piston Skirt Clearance	Refer to Paragraphs 104 and 105					
Crankshaft End Play	0.004	0.004	0.004	0.004	0.004	0.004
Camshaft Bearing Running Clearance	0.0025	0.0025	0.0025	0.0025	0.0025	0.0025
Cooling System—Gallons	3.75	3.75	3.75	3.75	3.75	3.75
Crankcase—Quarts (with Filter)	5	5	5	5	5	5
Transmission—Quarts (4-Speed)	6.5	6.5	6.5	6.5	6.5	6.5
(5-Speed)	8	8	8	8	8	8
"Select-O-Speed"		12		12	12	12
Differential—Quarts	8	8	(3)	(3)	8	(3)
Final Drive Housings, Each—Qts. (Row-Crop)	1.5	1.5	1.5	1.5	1.5	1.5
Hydraulic Reservoir—Quarts	8(4)	8(4)	8(4)	8(4)	8(4)	8(4)
Steering Gear Housing	1.5 lbs.	1.5 lbs.	1.5 lbs.	1.5 lbs.	1.5 lbs.	1.5 lbs.
Power Steering System	Fill to Proper Level — See Paragraph 37, 49, 60 or 68					

(1) 7.5:1 gasoline; 8.64:1 LP-Gas.

(2) For gasoline; 150-160 for LPG.

(3) Row Crop Models Only, 8 Quarts — All Other Models, 11½ Quarts.

(4) On models without both hydraulic system and PTO, leave hydraulic reservoir empty. On models without hydraulic system, but equipped with PTO, add six (6) quarts of oil to hydraulic reservoir.

FRONT SYSTEM AND MANUAL STEERING
ALL-PURPOSE TYPE

(LCG Type Front System and Manual Steering are Similar)

SPINDLE BUSHINGS

1. Refer to Fig FO1. To renew the spindle bushings, support front of tractor and disconnect the steering arms from the wheel spindles. Slide spindle and wheel assemblies out of axle extensions. Drive old bushings from axle extensions and install new ones using a piloted drift. New bushings will require no final sizing if not distorted during installation. Renew thrust bearing if unduly noisy.

AXLE CENTER MEMBER AND PIVOT PIN BUSHING

2. To remove the axle center member (18—Fig. FO1), support front of tractor, remove grille and unbolt radius rods and axle extensions from axle center member. Swing the axle extension and wheel assemblies away from tractor. On early models with non-threaded pivot pins, remove the cap screw retaining the axle pivot pin to the front end support and, using a slide hammer, remove the pivot pin. On models with threaded pivot pin, remove the cap screw (23—Fig. FO1) and locking flange (22) and unscrew the threaded pin by turning it counter-clockwise. Loosen radiator retaining nuts at bottom if necessary, to provide removal clearance and withdraw axle center member from either side of tractor.

The axle pivot pin bushing (17) can be renewed at this time. Make certain that pivot pin (20) has a free fit in the bushing before reinstalling the axle center member.

Install snap ring (21) on the late production threaded pin (20) on standard models, turn pin in until snap ring is tight against front support and then back pin out so that retainer (22) can be installed. On "heavy duty" axles with hex head pivot pin, tighten pin to a torque of 200 Ft.-Lbs. On all models, tighten radius rod to center axle bolts (16) to a torque of 75-135 Ft.-Lbs.

FRONT SUPPORT

3. To remove front support, remove grille on all models. Remove lower front panel and hood to front

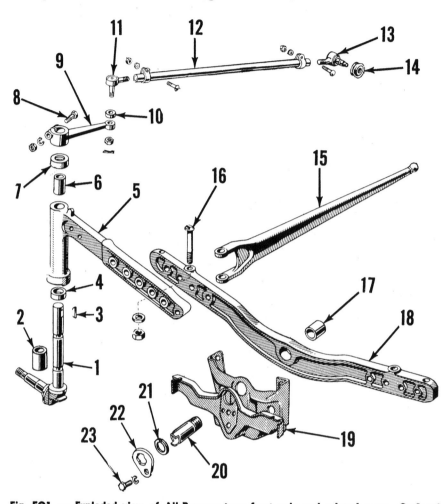

Fig. FO1 — Exploded view of All-Purpose type front axle and related parts. Optional "heavy duty" front axle is similarly constructed except that pivot pin (20) has hex head, snap ring (21) is not used and retainer (22) has hex opening for pivot pin head. A flat washer (not shown) is used on cap screw (23) between retainer (22) and front support (19) on "heavy duty" assemblies to prevent retainer from being cocked. Early production pivot pin was not threaded and pivot pin (20) and retainer (22) were integral welded assembly.

1. Spindle (R.H.)	7. Dust seal	13. Drag link end	19. Front support
2. Lower spindle bushing	8. Clamp bolt	14. Dust cover	20. Pivot pin
3. Woodruff key	9. Steering arm (R.H.)	15. Radius rod	21. Snap ring
4. Thrust bearing	10. Dust seal	16. Radius rod bolt	22. Retainer
5. Axle extension (R.H.)	11. Drag link end	17. Bushing	23. Cap screw
6. Upper spindle bushing	12. Drag link	18. Axle center member	

support bolts on 801 and 4000 Series. Unbolt radiator from front support. Place floor jack under front end of transmission, then remove axle pivot pin as in Paragraph 2. Remove nuts from front support to engine studs. Remove studs or pry front support forward to clear studs, then remove front support from below. Lower hood side panels will spring out far enough to clear lower part of radiator when moving front support forward.

When installing, tighten the retaining stud nuts to a torque of 135-150 Ft. Lbs.

DRAG LINKS AND TOE-IN

4. Drag link ends are of the non-adjustable automotive type. The procedure for renewing the drag link ends is evident. Vary the length of each drag link an equal amount to provide a front wheel toe-in of 1/4 to 1/2 inch.

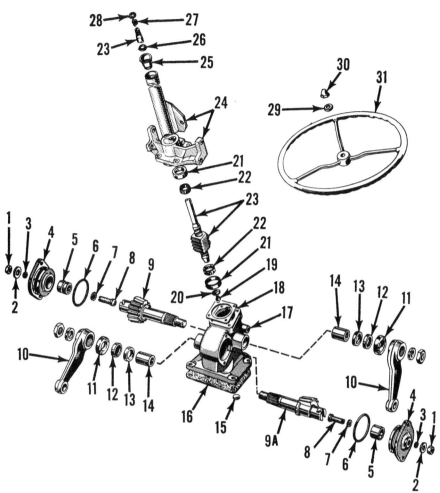

Fig. FO2 — Exploded view of manual steering gear assembly used on All-Purpose and LCG tractors. Wormshaft bearing end play is adjusted by varying the thickness of the shim stack (18); shims are available in thicknesses of 0.002, 0.005, 0.010 and 0.030. Sector end play is adjusted with adjusting screws (8). Use thickness of shim (7) that will provide zero to 0.002 clearance of adjusting screw head in slot of sector gear (9 or 9A).

1. Lock nuts	8. Adjusting screw	15. Expansion or Hex plug
2. Flat washers	9. Sector gear (double)	16. Gasket (Select-O-Speed
3. Packing	9A. Sector gear (single)	**only)**
4. End covers	10. Pitman arms	17. Gear housing
5. Bushings	11. Dust seals	18. Shims
6. "O" ring	12. Packing retainers	19. Eyelet
7. Shim (0.063, 0.065,	13. Packing	20. Bearing retainer
0.067 or 0.069)	14. Bushings	

21. Bearing cups	
22. Thrust bearings	
23. Wormshaft assembly	
24. Wormshaft housing	
25. Bearing	
26. Spring seat	
27. Coil spring	
28. Dust seal	

STEERING GEAR

NOTE: Late production industrial models are equipped with heavy duty steering sector gears (9 and 9A—Fig. FO2). Only the heavy duty sector gears will be available for servicing all models when current stocks of standard sector gears are exhausted. If necessary to renew a standard sector gear using a heavy duty gear, both sector gears must be renewed. Gears may be identified as follows: Standard sector gear (9A) has 7 full teeth; heavy duty sector gear (9A) has 5 full teeth. Standard sector gear (9) has 6 full teeth and 3 rack teeth; heavy duty sector gear (9) has 4 full teeth and 3 rack teeth. Service procedures remain unchanged.

5. **ADJUSTMENT.** To adjust the steering gear, first make certain that gear housing is properly filled with lubricant, disconnect both drag links from steering gear arms to remove load from the gear unit and proceed

as follows:

6. WORMSHAFT END PLAY. To check wormshaft end play, first loosen the lock nuts (1—Fig. FO2) on the sector shaft adjusting screws (8) and back the screws out at least two full turns. If the end play of the wormshaft (steering wheel shaft) is not within the desired limits of 0.006-0.010, adjust the end play by varying the thickness of the shim stack (18) between the steering shaft tube and the steering gear housing. Shims are available in thicknesses of 0.002, 0.005, 0.010 and 0.030. Ford recommends a minimum shim stack installation of not less than three 0.002 shims or not less than two 0.005 shims. Tighten the steering shaft cover retaining cap screws to a torque of 25-30 Ft.-Lbs. Renew wormshaft bearings as outlined in paragraph 10 if end play is over 0.010 with minimum recom-

mended shim stack thickness.

After checking or adjusting wormshaft end play, readjust sector shaft end play as follows:

7. SECTOR SHAFT END PLAY. Before adjusting sector shaft end play, be sure that wormshaft end play is correctly adjusted as outlined in paragraphs 5 and 6, then proceed as follows: Turn the steering wheel to the mid or straight ahead position. With the lock nuts on both sector shaft adjusting screws loosened and the adjusting screw on the **right** hand side (as viewed from rear of tractor) backed out several turns, turn the adjusting screw on **left** side of steering housing in (clockwise) until there is no perceptible end play in the sector shaft to which the **right** steering arm is attached. While holding the adjusting screw in this position, tighten the lock nut. Then, turn the adjusting screw on **right** side of housing in until there is no perceptible end play in the sector shaft to which the **left** steering arm is attached, hold the adjusting screw in this position and tighten the lock nut.

Reconnect the drag links to the steering arms.

8. **REMOVE AND REINSTALL.** To remove the steering gear and housing assembly, first remove steering wheel, then withdraw the spring, felt packing and spring seat from top of steering column. On "Select-O-Speed" transmission models, remove PTO control and gear selector as outlined in paragraph 234. Remove hood. Disconnect throttle rod from bell crank and unbolt throttle rod bracket from transmission. Disconnect the Proof-Meter cable, ammeter lead wire and oil pressure gage line at instrument panel. Disconnect the battery ground cable from steering gear housing and wires from junction block on steering column.

Remove the generator regulator from bracket on steering column and disconnect the temperature gage wire from fuel tank frame. Unbolt instrument panel, slide the panel over top of steering column and lay it on top of fuel tank. Unbolt battery carrier from steering gear housing. Disconnect head light switch and ignition switch from the hood rear lower panel; then, unbolt and remove the hood rear lower panel from tractor. Unclip tail light wire from steering gear housing and disconnect drag links from pitman arms. Remove the cap screws retaining steering gear housing to transmission case and lift the steering gear assembly from tractor. NOTE: On "Select-O-Speed" transmission models, a gasket is used between steering gear and transmission housings. Gasket should be left in place or opening in transmission covered when steering gear housing is removed. Be sure that gasket is in good condition before reinstalling steering gear assembly.

9. **OVERHAUL.** Major overhaul of the steering gear unit necessitates the removal of the unit from tractor as outlined in paragraph 8. Remove the pitman arm retaining nuts and pull pitman arms from sector shafts. Unbolt the sector shaft side covers and remove the adjusting screw lock nuts (1—Fig. FO2). Using a screwdriver, turn the adjusting screws in and remove the side covers and sector shafts. Unbolt steering housing upper cover from housing and remove cover, shaft and ball nut assembly. Do not disassemble the ball nut and steering shaft assembly (23) as component replacement parts are not available. If the steering shaft and/or ball nut are damaged, renew the complete assembly. The need and procedure for further disassembly and/or overhaul is self-evident.

The renewable bushings in steering gear covers have a bore diameter of 1.1255-1.1260; bushings in housing have a bore diameter of 1.245-1.250.

Shims (7) on the adjusting screws (8) are available in thicknesses of 0.063, 0.065, 0.067 and 0.069. When reassembling, use a shim that will provide zero to 0.002 clearance between adjusting screw head and slot in sector shafts.

When reassembling, center the ball nut on wormshaft and insert shaft in housing. Bolt the housing upper cover assembly in position, using the necessary number of shims (18) to provide an end play of 0.006-0.010 of wormshaft in bearing. Minimum shim stack should be three 0.002 shims or two 0.005 shims. If end play is more than 0.010 with minimum recommended thickness of shims, renew the worm-shaft bearings. Shims (18) are available in thicknesses of 0.002, 0.005, 0.010 and 0.030. Tighten the cover cap screws to a torque of 25-30 Ft.-Lbs. Assemble the sectors and their adjusting screws (8) to their covers. Hold the left sector shaft (the one with the greater number of teeth) and side cover assembly with the block tooth up, and install the sector so that middle tooth on sector meshes with middle groove on the ball nut rack. Install the right sector shaft, meshing the fourth tooth with the fourth groove of the left sector shaft. Tighten the side cover cap screws to a torque of 25-30 Ft.-Lbs. and install the adjusting screw lock nuts.

Turn steering gear to its mid or straight ahead position and install pitman arms.

When installation is complete, fill gear housing with lubricant and adjust the sector shaft end play as outlined in paragraph 7. Reconnect the drag link.

FRONT SYSTEM, UTILITY INDUSTRIAL TYPE

(Front System of Grove Type is Similar. Utility Industrial and Grove Types Are Equipped With Power Steering Only.)

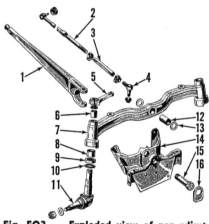

Fig. FO3 — Exploded view of non-adjustable front axle and associated parts used on Utility Industrial and Grove tractors.

1. Radius rod	9. Bearing
2. Drag link	10. Washer
3. Adjusting sleeve	11. Spindle
4. Tie rod end	12. Bushing
5. Steering arm	13. Washer
6. Upper bushing	14. Front support
7. Axle	15. Pivot pin
8. Lower bushing	16. Locking flange

SPINDLE BUSHINGS

10. Refer to Fig. FO3. To renew the spindle bushings (6 and 8), support front of tractor, disconnect steering arms (5) from wheel spindles (11) and slide wheel and spindle assemblies out of axle (7). Drive old bushings from axle and install new bushings using a piloted drift. New bushings will not require final sizing if not distorted during installation. Renew thrust bearings (9) if rough or worn.

FRONT AXLE AND PIVOT PIN

11. To remove the front axle, remove hood and radiator; then support front of tractor. Refer to Fig. FO3. If front axle is to be renewed, disconnect steering arms (5) from spindles (11) and withdraw spindles downward out of axle assembly.

Remove the locking flange (16), and unscrew the threaded pin (15). Remove the radius rods (1) and lift front axle (7) from front support.

Pivot bushing (12) can be installed without removing axle, by removing pivot pin (15) and allowing tractor to drop until axle clears front support. Bushing is not a tight fit in axle.

When reassembling, tighten the pivot pin (15) to a torque of 200 Ft.-Lbs.; and the radius rod to axle bolts to a torque of 75-135 Ft.-Lbs.

FRONT SUPPORT

12. To remove front support, remove radiator grille on all models. Remove lower front panel on 4000 Series. Unbolt radiator from front support. Place floor jack under front edge of transmission; then, remove pivot pin as in paragraph 11. Remove nuts from front support to engine studs. Remove studs or pry front support forward to clear studs, then remove front support from below. Lower hood side panels will spring out far enough to clear lower part of radiator when moving front support forward.

When installing, tighten the front support retaining stud nuts to a torque of 135-150 Ft.-Lbs.

DRAG LINKS AND TOE-IN

13. Drag link ends are of the non-adjustable automotive type. The procedure for renewing the drag link ends is evident. Vary the length of each drag link an equal amount to provide a front wheel toe-in of ¼ to ½-inch.

STEERING GEAR

All Utility Industrial models and Grove type tractors are equipped with power steering. The steering gear is an integral part of the power steering unit. Refer to paragraph 40.

FRONT SYSTEM, H. D. INDUSTRIAL TYPE

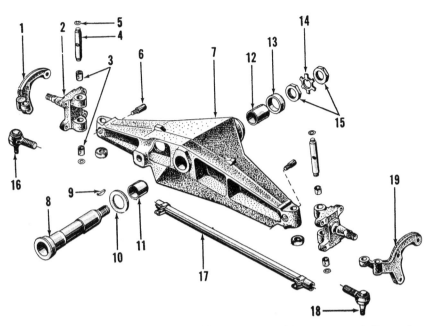

Fig. FO4 — Exploded view (from rear) of front axle assembly used on early Heavy Duty Industrial tractor with full power steering.

1. Left steering arm	6. Tapered retaining pin	11. Rear pivot bushing	16. Tie rod end
2. Spindle	7. Axle member	12. Front pivot bushing	17. Tie rod
3. Spindle bushing	8. Pivot pin	13. Thrust bearing	18. Tie rod end
4. Spindle pin	9. Woodruff key	14. Tab washer	19. Right steering
5. Pin seal	10. Thrust washer	15. Jam nut	arm

SPINDLE BUSHINGS

All Models

14. To renew spindle bushings, jack up the front axle and remove wheel and hub assembly. Cut lock wire or bend locking tabs down and unbolt steering arm from spindle. Tie-rod and/or power steering cylinder need not be disconnected from steering arm. Remove nut and lock washer from tapered lock pin (6—Fig. FO4 or Fig. FO5) and drive out tapered pin. Drive spindle pin upward enough to relieve the pressure on upper spindle pin seal, remove upper seal, then drive spindle pin downward out of axle.

Spindle bushings are pre-sized and will require no final sizing if carefully installed. Renew thrust bearing if unduly noisy. Thrust bearing is installed with the indentation up as shown in Fig. FO7. The detent in the spindle pin is off center. Install spindle pin with stamped "T" mark facing upward so that spindle pin will be properly located in axle. Torque steering arm bolts to 100 Ft.-Lbs., and secure with safety wire or locking tabs.

FRONT AXLE AND PIVOT PIN

15. To remove the front axle (7—Fig. FO4 or 7A—Fig. FO5), first disconnect the tie rod from either steering arm and swing tie rod rearward. Disconnect power cylinder pressure and return lines at cylinder; then, disconnect the power cylinder from the axle and lower the cylinder assembly to the floor. Remove the radiator grille door, pass a chain loop through hole of front axle support and secure chain with a bolt or short rod as shown in Fig. FO6. Attach a hoist to chain and tighten enough to take up slack. Place a rolling floor jack under center of front axle and raise jack just enough to take the axle weight off of pivot pin. Straighten tabs on keyed lock washer at front end of pivot pin and remove jam nut, keyed washer and lock nut as

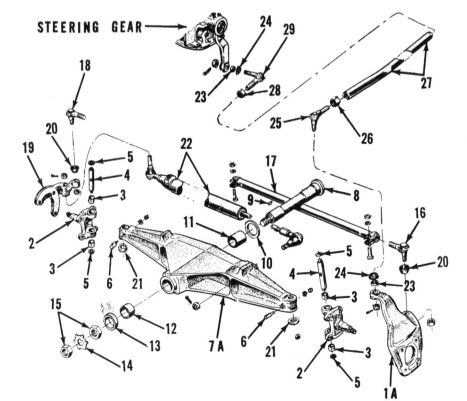

STEERING GEAR ——→

Fig. FO5 — Exploded view of late production Heavy Duty Industrial axle assembly with drag link (power assist) steering. View is from front of unit.

1A. Left steering arm	16. Tie rod end
2. Spindles	17. Tie rod
3. Spindle bushings	18. Tie rod end
4. Spindle pins	19. Right steering arm
5. Spindle pin seals	20. Dust cover
6. Tapered retaining pins	21. Thrust bearings
7A. Front axle	22. Power steering
8. Pivot pin	cylinder
9. Woodruff key	23. Dust cover
10. Thrust washer	24. Retainer
11. Rear pivot bushing	25. Drag link end
12. Front pivot bushing	26. Lock nut
13. Thrust spacer	27. Drag link
14. Tab washer	28. Lock nut
15. Jam nuts	29. Drag link end

9

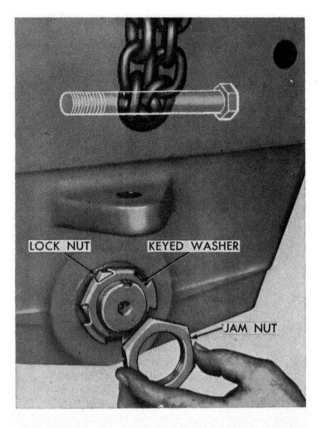

Fig. FO6 — Support front of Heavy Duty Industrial tractor with chain loop and bolt, as shown, when removing front axle or pivot pin. Note method of locking front end of axle pivot pin.

LOCK NUT   KEYED WASHER   JAM NUT

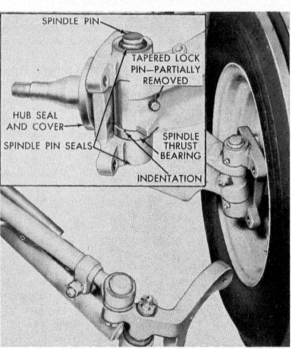

Fig. FO7 — Spindle on Heavy Duty Industrial attaches to axle by means of spindle pin and tapered lock pin as shown. Install spindle thrust bearing with indentation up.

SPINDLE PIN — TAPERED LOCK PIN—PARTIALLY REMOVED — HUB SEAL AND COVER — SPINDLE PIN SEALS — SPINDLE THRUST BEARING — INDENTATION

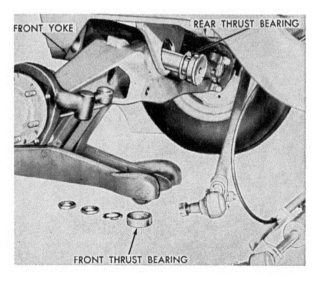

Fig. FO8 — Tilt front axle to the rear as shown, when removing pivot pin from Heavy Duty Industrial tractor. Left wheel is removed for illustration purposes only. (Early model with full power steering is shown.)

FRONT YOKE   REAR THRUST BEARING   FRONT THRUST BEARING

axle assembly and move axle forward away from tractor as shown in Fig. FO9.

Renew the two bushings in the front axle as required. The diameter of the two bushings is the same but the front bushing is longer. Drive the bushings into the axle member until ends are flush with axle.

Reinstall by reversing the removal procedure, keeping in mind that the Woodruff key must be aligned with the keyway in the front bore of the support. Tighten pivot pin lock nut to 50 Ft.-Lbs. and secure with tab washer and jam nut.

FRONT SUPPORT

16. To remove the front axle support, first remove the radiator as described in paragraph 152. Disconnect the hydraulic lines from the power package pump and cap all exposed openings. Remove the four nuts from the pump mounting plate studs and pull the pump, mounting plate and universal drive shaft as a unit from the tractor. Place a floor jack under the transmission case and support the weight of the tractor. Position a second jack under the center of the front axle to stabilize the axle. Remove the bolts securing the support to the side members and the two bolts securing the support to the front of the engine. Disconnect the power steering cylinder hoses and unbolt and remove the cylinder. Slide the support and front axle assembly forward until the axle pivot pin will clear the engine and remove the pivot pin. Raise the front of the tractor enough for the front support to clear the front axle and slide the support forward out of the side rails.

shown. With all tension removed from pivot pin, drift pin rearward until it clears front yoke of the axle support, then remove front thrust bearing. Reposition floor jack to tilt axle rearward as shown in Fig. FO8 enough to allow pivot pin, rear thrust bearing and Woodruff key to be removed from the rear. Raise front of tractor with hoist enough to clear

DRAG LINK, TIE ROD
AND TOE-IN

17. On models with power assist steering, the installed length of the drag link should be adjusted to $46\frac{7}{16}$ to $46\frac{1}{2}$ inches, measured between centers of sockets at each end of the drag link. The offset (elbow) in the drag link must be towards the front end of the tractor, and point 15 to 20 degrees below horizontal and away from the engine. If tractor is equipped with loader, check clearance between drag link and left-hand cylinder on loader in all positions of front axle, loader and drag link. If interference exists, loosen lock nuts at each end of drag link, rotate sleeve slightly to provide clearance and retighten lock nuts.

18. The tie rod ends (and drag link ends on power-assist steering models) are of the non-adjustable automotive type. Renewal procedure is evident.

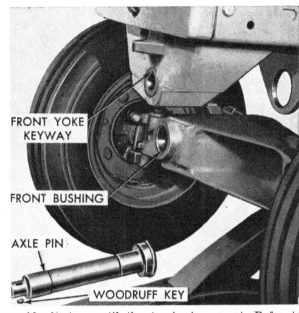

Fig. FO9 — Raise front of Heavy Duty Industrial tractor with a hoist and roll front axle and wheels assembly forward away from tractor.

FRONT YOKE KEYWAY

FRONT BUSHING

AXLE PIN

WOODRUFF KEY

Adjust the tie rod to provide ¼ to ½-inch toe-in by loosening the clamp on each end and rotating the tube until the toe-in is correct. Refer to paragraph 17 for adjustment of drag link.

FRONT SYSTEM AND STEERING GEAR, ROW CROP

PEDESTAL AND COMPONENTS
Early Series 700-900

19. **REMOVE AND REINSTALL.** To remove the pedestal, proceed as follows: Provide support for the front of the tractor and drain the radiator. Remove the bolts retaining hood to pedestal and the nuts from the radiator retaining studs. Disconnect drag link from front steering arm. Support pedestal in a suitable manner, unbolt pedestal from engine and side rails and move the pedestal and wheels assembly away from the tractor. The radiator will be supported by the radiator hoses and the fan after the pedestal is removed. See Fig. FO10.

Note: It may be necessary to loosen the generator adjusting bracket and the lower radiator hose to gain access to one of the pedestal lower retaining bolts.

Replacement pedestals are factory fitted with bushings, bearing cup and seals.

20. **OVERHAUL.** Normal overhaul of the pedestal can be accomplished without removing the pedestal from the tractor as follows:

Remove grille, drain the pedestal oil reservoir and remove the reservoir cover. Raise the front of the tractor until the bottom of the pedestal is approximately 18 inches from the floor and remove the front wheel and hub assemblies. Remove the front steering

arm (4—Fig. FO10), thrust cap (5) and shim stack being careful not to damage or lose the shims. Support the vertical spindle (18), remove nut from top of spindle shaft and bump the vertical spindle down and out of pedestal. Turn steering sector arm shaft (7) until the bevel gears unmesh and withdraw gear (11). Pull

steering sector arm shaft (7) and gear unit out through top opening in pedestal. The need and procedure for further disassembly is self-evident.

Bushings (1) for sector arm shaft and bushing (10) at top of spindle are renewable and require no final sizing if carefully installed using a suitable piloted driver. Renewal of

Fig. FO10 — Exploded view of early 700 and 900 Series Row Crop pedestal. Design of pedestal was changed in 1957 production to that shown in Fig. FO11 to allow installation of single front wheel or wide adjustable front axle.

1. Bushings
2. Shims
3. Oil seal
4. Steering arm
5. Thrust cap
6. Pedestal
7. Sector shaft
8. Oil seal spacer
9. Oil seal
10. Bushing
11. Sector gear
12. Cover
13. Oil dipstick
14. Gasket
15. Bearing cup
16. Cone & roller assy.
17. Dust seal
18. Spindle
19. Drain plug

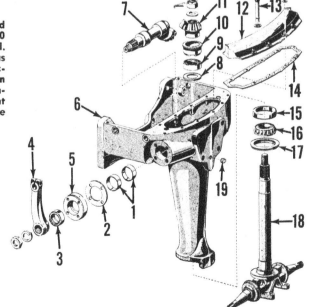

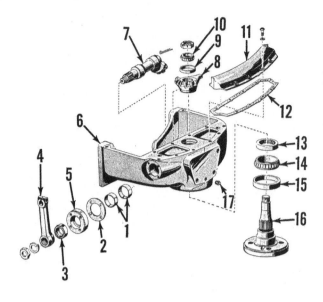

Fig. FO11 — Exploded view of pedestal used on late production 700 and 900 Series and all other later model Row Crop tractors. Wide adjustable front axle (Fig. FO13), dual wheel (Fig. FO14) or single wheel spindle (Fig. FO15) can be installed on this type pedestal.

1. Bushings
2. Shims
3. Oil seal
4. Steering arm
5. Thrust cap
6. Pedestal
7. Sector shaft
8. Sector gear
9. Bearing cup
10. Cone & roller assy.
11. Cover
12. Gasket
13. Oil seal
14. Cone & roller assy.
15. Bearing cup
16. Spindle
17. Drain plug

PEDESTAL AND COMPONENTS

Late Series 700 and 900, All Other Later Rowcrop Models

21. REMOVE AND REINSTALL. The procedure for removing and re-installing the pedestal is the same as the procedure outlined in paragraph 19.

22. OVERHAUL. Normal overhaul of the pedestal can be accomplished without removing the pedestal from the tractor as follows: (Refer to Fig. FO11).

Remove grille, drain the pedestal oil reservoir and remove the pedestal cover. Raise front of tractor and on tricycle models, unbolt and remove the wheel spindle from the vertical spindle shaft. On adjustable axle models, unbolt the center steering arm from shaft spindle and axle support from pedestal. Roll the axle and wheels assembly forward and away from tractor. Remove the steering arm, thrust cap and shim stack from sector shaft and be careful not to damage or lose the shims. Remove nut from top of vertical spindle shaft, and bump the spindle shaft down and out of pedestal. Turn the sector shaft until gears unmesh and remove the sector gear. Withdraw the sector shaft and gear unit out through top of pedestal. The need and procedure for further disassembly is evident.

Sector shaft bushings are renewable and require no final sizing if carefully installed, using a piloted driver. Spindle shaft oil seal should be pressed into pedestal until bottom of seal is 1½ inches from the machined bottom face of pedestal as shown in Fig. FO12.

Install the sector shaft and sector gear, making certain that gears are meshed at the center of the gear travel so that front wheels will turn the same distance in both directions. Install the spindle shaft bearing cups and cones. Pack the lower tapered roller bearing (14—Fig. FO11) with Ford M-4664 or equivalent high melting point grease. Then install the vertical spindle so that master splines engage master splines in sector gear. Install nut on top of spindle, tighten the nut to remove all spindle end play without causing any drag and install the cotter pin.

Install the original shim stack and lubricate lips of seal with grease. Install sector shaft cap and tighten the cap retaining screws to a torque of

Fig. FO12 — On series 700 and 900 produced after 1956, and all other Row Crop models, the spindle oil seal should be positioned 1½ inches from the machined lower face of the pedestal.

spindle upper oil seal (9) necessitates removal of spindle bushing. Spindle lower dust seal (17) seats in bottom of pedestal. When reassembling, position spacer (8) in the pedestal and using Nuday Tool No. N-506 or equivalent, drive the seal (9) into the pedestal until the seal bottoms against the spacer. Reinstall the sector shaft and sector gear, making certain that gears are meshed at the center of the gear travel so that front wheels will turn the same distance in both directions. Reinstall the vertical spindle so that master splines engage master splines in gear (11). Install washer with flat side down and

tighten nut to remove all spindle end play without causing any drag.

Install the original shim stack (2), thrust cap (5) and tighten the retaining cap screws to a torque of 25-30 Ft.-Lbs. With the front wheels pointing straight ahead, there should be a backlash of 0.0005-0.0035 between the bevel gears. If the backlash is not as specified, vary the number of shims (2) which are available in thicknesses of 0.0025, 0.007, 0.010 and 0.020.

Complete the assembly by reversing the disassembly procedure and fill pedestal to proper level with SAE 80 EP gear lubricant.

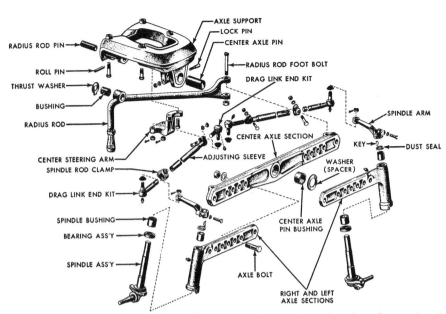

Fig. FO13 — Exploded view of the adjustable type front axle used on Row Crop pedestal shown in Fig. FO11. Axle and radius rod pivot bushings are pre-sized.

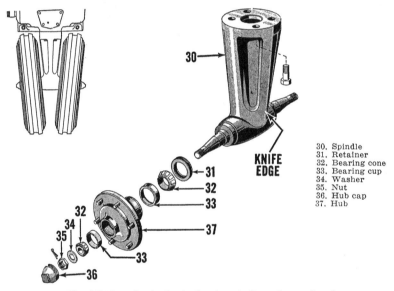

Fig. FO14 — Dual wheel tricycle spindle and associated parts used on Row Crop pedestal shown in Fig. FO11. Knife edge of spindle goes toward front of tractor.

30. Spindle
31. Retainer
32. Bearing cone
33. Bearing cup
34. Washer
35. Nut
36. Hub cap
37. Hub

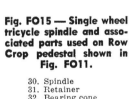

Fig. FO15 — Single wheel tricycle spindle and associated parts used on Row Crop pedestal shown in Fig. FO11.

30. Spindle
31. Retainer
32. Bearing cone
33. Bearing cup
34. Washer
35. Nut
36. Hub cap
38. Wheel disc
39. Wheel flange

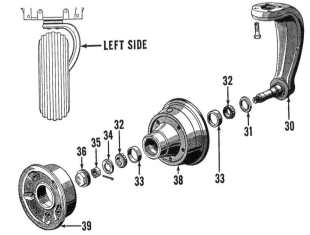

35-40 Ft.-Lbs. Note: Late production thrust cap was reduced in thickness from 7/8-inch to 5/16-inch. Also a new seal assembly was incorporated which has a 3½-inch flange and is bolted to the thrust cap instead of being pressed into a recess. With the front wheels pointing straight ahead, there should be a backlash of 0.001-0.004 between the bevel gears. If the backlash is not as specified, vary the number of shims which are available in thicknesses of 0.0025, 0.007, 0.010 and 0.020.

Complete the assembly by reversing the disassembly procedure and install two quarts of SAE 80 EP gear lubricant in the pedestal. Steering arm should be in a vertical position when front wheels are pointing straight ahead.

WIDE ADJUSTABLE ROWCROP FRONT AXLE

23. SPINDLE BUSHINGS. An exploded view of the adjustable type front axle is shown in Fig. FO13. The procedure for renewing the spindle bushings is conventional and evident after an examination of the unit.

24. AXLE, RADIUS ROD, PIVOT PINS AND BUSHINGS. The easiest and quickest way to remove the center member, radius rod, pivot pins and/or bushings is to support front half of tractor, unbolt center arm from vertical spindle and axle support from pedestal; then roll the axle, support and wheels assembly forward and away from tractor. Remove the taper pin and roll pin retaining the axle and radius rod pivot pins in the axle support, drift out the pins and lift off the axle support. The need and procedure for further disassembly is evident.

Pivot bushings in axle and radius rod are renewable and will not require final sizing if carefully installed.

When reassembling, tighten the radius rod foot bolts to a torque of 150-180 Ft.-Lbs. Be sure to install the same number of spacer washers at front of axle center section as were originally removed. Thrust washer at rear of radius rod should be installed with hooked projection down and toward rear of tractor.

25. DRAG LINKS AND TOE-IN. Drag link ends are of the non-adjustable automotive type. The procedure for renewing the drag link ends is evident. Vary the length of each drag link an equal amount to provide a front wheel toe-in of ⅛-¼ inch.

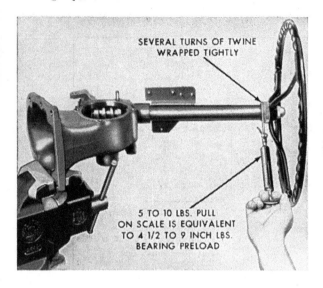

Fig. FO16 — Method of checking preload on Row Crop steering wormshaft bearings. Note where twine is wrapped.

SEVERAL TURNS OF TWINE WRAPPED TIGHTLY

5 TO 10 LBS. PULL ON SCALE IS EQUIVALENT TO 4 1/2 TO 9 INCH LBS. BEARING PRELOAD

MANUAL STEERING GEAR

All Rowcrop Models

Note: This section applies only to the gear unit mounted on transmission. For overhaul of the bevel gears located in pedestal, refer to paragraph 20 or 22.

Early 700 and 900 Series were equipped with manual steering; however, most have been converted to power steering which is standard equipment on all later rowcrop models. Same steering gear is used for both manual and power steering.

26. ADJUSTMENT. Before attempting to adjust the steering gear, first disconnect the drag link from pitman arm and back off the adjustments as follows: Tap the lash adjuster lock plate counter-clockwise (opposite the direction of the arrow stamped on plate) after loosening the lock bolt (31—Fig. FO18). Turn the sector shaft end play adjusting screw (9) counter-clockwise after loosening lock nut.

27. WORMSHAFT BEARINGS. First step is to back off the other adjustments and disconnect drag link as outlined in paragraph 26. Remaining procedure is as follows: Alternately pull up and push down on steering wheel and note amount of end play in wormshaft. If end play is greater than 0.002 or if shaft binds when rotated, loosen the locknut (30) and turn the end play adjusting screw (25) until zero end play without binding, is obtained. The recommended adjustment is 5-10 pounds pull to rotate shaft as shown in Fig. FO16. This recommendation however applies only when test is made with sector shaft removed from gear housing as shown.

28. SECTOR SHAFT END PLAY. With worm shaft bearings adjusted as in paragraph 27, turn the adjusting screw (9—Fig. FO18) clockwise to remove all end play from sector shaft without causing any binding tendency.

29. SECTOR BACKLASH. To adjust the tooth backlash, adjust the other points on the gear as outlined in paragraphs 26, 27 and 28, then proceed as follows: Turn steering wheel to mid-position (center) of its travel and tap the adjusting plate (29) clockwise (in the direction of arrow stamped on plate) until all backlash is removed from the sector shaft. The recommended adjustment is 16-26 pounds pull of a spring scale

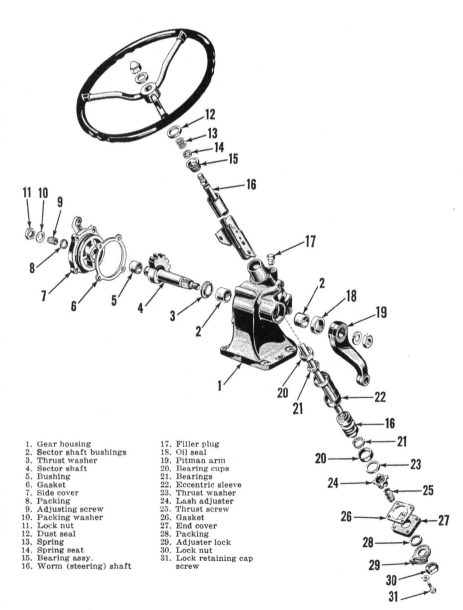

1. Gear housing	17. Filler plug
2. Sector shaft bushings	18. Oil seal
3. Thrust washer	19. Pitman arm
4. Sector shaft	20. Bearing cups
5. Bushing	21. Bearings
6. Gasket	22. Eccentric sleeve
7. Side cover	23. Thrust washer
8. Packing	24. Lash adjuster
9. Adjusting screw	25. Thrust screw
10. Packing washer	26. Gasket
11. Lock nut	27. End cover
12. Dust seal	28. Packing
13. Spring	29. Adjuster lock
14. Spring seat	30. Lock nut
15. Bearing assy.	31. Lock retaining cap
16. Worm (steering) shaft	screw

Fig. FO17 — Exploded view of Row Crop steering gear. Same unit is used for both manual and power steering.

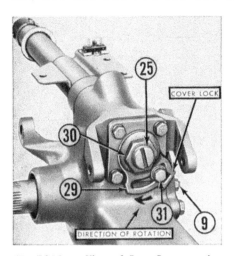

Fig. FO18 — View of Row Crop steering housing end cover showing lash adjuster lock plate (29) and end play (thrust) adjusting screw (25). Sector shaft end play is controlled by screw (9).

Fig. FO19 — View showing eccentric sleeve and splines of lash adjuster which must engage notches in the eccentric sleeve.

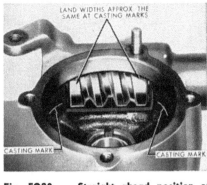

Fig. FO20 — Straight ahead position of steering worm will be obtained when land widths are equal at casting marks as shown.

Fig. FO21 — When middle tooth of sector is meshed with middle groove of worm, the sector is properly positioned in main steering gear unit.

connected to wormshaft as shown in Fig. FO16. This amount of pull should register just as the gear passes through the mid (center) position and will be less on either side of center.

30. **REMOVE AND REINSTALL.** To remove the steering gear and housing assembly, first remove the steering wheel, then withdraw the spring, felt packing and spring seat from top of steering column. On "Select-O-Speed" transmission models, remove the PTO control and gear selector assemblies as outlined in paragraph 234. Disconnect throttle rod from bell crank and unbolt throttle rod bracket from transmission. Disconnect the Proof-Meter cable, ammeter lead and oil pressure gage line at instrument panel. Disconnect the battery ground cable from steering gear housing and wires from junction block on steering column. Remove the generator regulator from bracket on steering column and disconnect the temperature gage tube from gas tank frame.

Unbolt instrument panel, slide the panel over top of steering column and lay it on top of fuel tank. Unbolt battery carrier from steering gear housing and disconnect head light switch and ignition switch from the hood rear lower panel. Unbolt and remove the hood panel from tractor. Unclip tail light wire from steering gear housing and disconnect drag link from pitman arm. Remove the cap screws retaining steering gear hous-

ing to transmission case and lift the steering gear assembly from tractor. NOTE: On "Select-O-Speed" transmission models, a gasket is used between the steering gear and transmission housings. Gasket should be left in place when steering gear is removed or opening in transmission covered to keep foreign material from entering transmission and causing serious damage. Be sure gasket is in good condition before reinstalling steering gear assembly.

31. **OVERHAUL.** Major overhaul of the steering gear unit necessitates the removal of the unit from tractor as in paragraph 30. Remove the pitman arm retaining nut and pull pitman arm from sector shaft. Unbolt and remove side cover (7—Fig. FO17) and withdraw sector shaft and thrust washer (3) from housing. Remove end cover (27) and withdraw worm shaft and eccentric sleeve assembly (22) from housing. The need and procedure for further disassembly is evident.

Thoroughly clean and examine all parts for damage or wear. New sector shaft bushings (2) should be installed with a piloted drift. Install worm shaft, bearings and eccentric sleeve assembly in the housing and position the sleeve so that cut-out in sleeve will allow sector gear teeth to mesh with the worm. The approximate position of the sleeve is shown in Fig.

FO19. Place the thrust washer (23—Fig. FO17) in the eccentric sleeve and assemble the end cap. Install the end cap assembly, making certain that tabs on the lash adjuster plate engage notches in the eccentric sleeve. Adjust the worm bearings as outlined in paragraph 27.

Position the steering worm shaft in the straight ahead position, as follows: Turn the shaft until the end land widths on the worm are approximately the same at the center casting marks in the housing as shown in Fig. FO20. Install the sector shaft and thrust washer, making certain that middle tooth of sector engages center groove of worm as shown in Fig. FO21. Install the housing side cover and adjust the sector shaft end play as in paragraph 28.

Adjust the gear unit backlash as outlined in paragraph 29 and install the gear unit by reversing the removal procedure.

FRONT SYSTEM & MANUAL STEERING, OFFSET

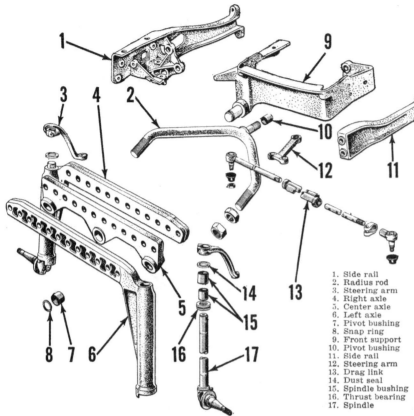

1. Side rail
2. Radius rod
3. Steering arm
4. Right axle
5. Center axle
6. Left axle
7. Pivot bushing
8. Snap ring
9. Front support
10. Pivot bushing
11. Side rail
12. Steering arm
13. Drag link
14. Dust seal
15. Spindle bushing
16. Thrust bearing
17. Spindle

Fig. FO22 — Front axle and support assembly used on Offset type tractors.

Fig. FO23 — To remove front axle assembly on Offset models, disconnect steering U-joint (1) and steering arm (2), remove steering gear housing mounting bolts (3) and lift gear housing upward out of support. Nut (4) secures radius rod to support.

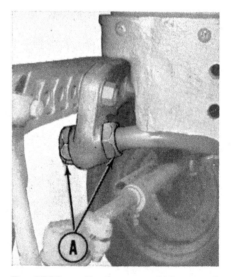

Fig. FO24 — Use jam nuts (A) at each end of radius rod to properly align front axle center member. See text.

SPINDLE BUSHINGS

32. Refer to Fig. FO22. To renew the spindle bushings (15), support front of tractor, disconnect steering arms (3) from wheel spindles (17) and slide wheel and spindle assemblies out of axle extensions (4 and 6). Drive old bushings from axle extensions and install new bushings using a piloted drift. New bushings will not require final sizing if not distorted during installation. Renew thrust bearings (16) if rough or worn.

AXLE CENTER MEMBER AND PIVOT PIN BUSHING

33. Refer to Fig. FO22 for an exploded view of the axle assembly, radius rod and front support. The assembly pictured is for the standard clearance models. On high clearance models, the axle center section is inverted, with pivot pin and radius rod bosses to top of axle. Different drag links and steering arms are used. To remove the front axle center section, support the tractor under the transmission case, disconnect the front universal joint (1—Fig. FO23) from wormshaft, and steering arm (2) from the pitman arm; then, unbolt and remove the steering gear housing assembly from front support. Remove cotter pin, nut (4) and washer from radius rod and extract snap ring (8—Fig. FO22) from the axle pivot pin. Move the entire axle assembly forward away from the tractor. The axle and radius rod pivot bushings (7 and 10) may be renewed at this time if worn or damaged.

When the front axle assembly has been disassembled for renewal of parts, the radius rod must be realigned to the axle center section as follows:

Reinstall the unit on the tractor, leaving the radius rod adjusting nuts (A—Fig. FO24) loose. Adjust the radius rod pivot nut (4—Fig. FO23) to remove end play and install cotter pin. Install snap ring at front of axle pivot pin and tighten rear adjusting nuts evenly against axle center member until front of axle just touches the snap ring and is square with the pin; then, tighten the two nuts at the front of the axle center member to lock the adjustment.

FRONT SUPPORT

34. To remove the front support assembly, first drain the radiator, remove the three bolts securing the rear steering column support to the transmission housing, disconnect the front steering column universal joint from the wormshaft and remove the steering column assembly. Disconnect the light wires and remove the hood, side

panels and grille assembly; then, unbolt and remove the radiator. Remove the front axle assembly as outlined in paragraph 33, then unbolt and remove the front support.

DRAG LINKS AND TOE-IN

34A. Drag link ends are of the non-adjustable automotive type. The procedure for renewing the drag link ends is evident. Vary the length of each drag link an equal amount to provide a front wheel toe-in of ¼-inch.

MANUAL STEERING GEAR

35. **ADJUSTMENT.** Backlash of the worm and sector type steering gear can be adjusted without disassembly. To adjust the backlash, place the front wheels in a straight ahead position, loosen the lock nut on the adjusting screw in gear cover which is accessible at the top right side of the front support. Turn adjusting screw (4—Fig. FO25) until all backlash is

eliminated. Worm shaft bearings are shim adjusted. To make this adjustment, unit must be removed as outlined in paragraph 36.

36. **R&R AND OVERHAUL.** To remove the steering gear assembly, disconnect the front universal joint from worm shaft and steering arm from the sector shaft. Remove the screws retaining steering housing to right side rail and withdraw the steering gear. Refer to Fig. FO25.

Bushings (6) are renewable. Be sure to remove all rust, burrs or paint from exposed end of worm and sector shafts before removing shafts, to prevent damage to the shaft seals and bushings. Examine shaft seals and renew if indicated.

Shims (9) control the end play worm shaft bearings (10 & 11). When assembling, add or remove shims as required, to remove all end play without binding. Shims are available in thicknesses of 0.002, 0.005 and 0.010.

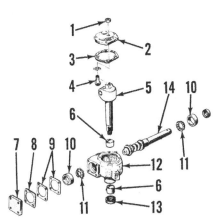

Fig. FO25 — Exploded view of worm and sector type steering gear used on Offset tractors. Steering gear attaches to inside front of right side rail and is controlled by a shaft and universal joint assembly.

1. Locknut	8. Gasket
2. Top cover	9. Shim stack
3. Gasket	10. Bearing cup
4. Adjusting screw	11. Bearing cone
5. Sector shaft	12. Housing
6. Shaft bushing	13. Oil seal
7. Front cover	14. Worm shaft

POWER STEERING SYSTEM

(All-Purpose, LCG and Utility Industrial Types. Grove Type is Similar)

The power steering system used on these tractors is of the linkage booster type which utilizes a control valve assembly combined integrally with the steering shaft and gear assembly. Refer to Fig. FO26.

The maintenance of absolute cleanliness of all parts is of utmost importance in the operation and servicing of the hydraulic power steering system. Of equal importance is the avoidance of nicks or burrs on any of the working parts.

LUBRICATION AND BLEEDING

37. Recommended fluid is Ford M2C41 hydraulic oil. Oil should be changed only when it is necessary to drain oil for repairs. Filter element should be renewed if damaged or if a considerable amount of sediment is noted.

Eight pressure type grease fittings (2 each on drag links and power cylinders) are provided and should be lubricated with pressure gun grease daily or after every 10 hours.

To bleed system, fill reservoir to full mark as indicated on dip stick on early all-purpose models with hollow plunger type pump. On roller vane type pump, remove reservoir cover and fill reservoir to about ⅝-inch from top. Start engine and

cycle gear full left and full right several times to bleed air from system. Recheck fluid level and refill if necessary.

sary. Be sure reservoir cover is installed securely on roller vane type pumps.

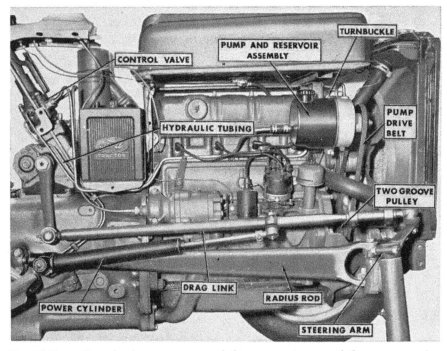

Fig. FO26 — View showing arrangement of the component parts of the power steering system used on early All-Purpose type tractors. Systems used on late All-Purpose type, Utility Industrial type, LCG type and Grove type are similar.

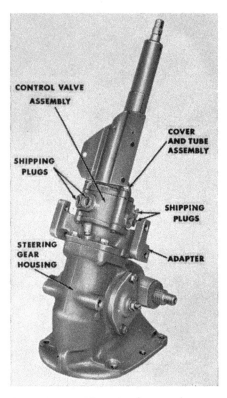

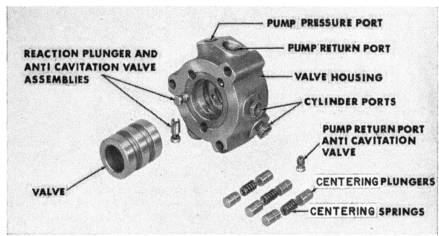

Fig. FO28 — Exploded view of control valve assembly. Valve and valve body are available only as a matched set.

Fig. FO27 — View showing steering gear and control valve assemblies removed from tractor. Note the shipping plugs installed in all ports.

SYSTEM OPERATING PRESSURE

38. A pressure test of the power steering circuit will disclose whether the pump, relief valve or some other unit in the system is malfunctioning. To make such a test proceed as follows:

Connect a pressure test gage in series with the pump pressure line. Run the engine at slow idle speed until the working fluid is warmed to normal operating temperature. Advance the engine speed to high idle rpm, turn the steering wheel to either side against stop and hold in this position only long enough to observe the gage reading. NOTE: Pump may be seriously damaged if steering wheel is held in this position for more than a few seconds. If the gage reading is 700-800 psi, the pump and relief valve are O. K. and any trouble is located in the control valve, power cylinders and/or connections.

If the gage pressure is more than 800 psi, the pump relief valve is probably stuck in the closed position, incorrectly installed, or wrong relief valve assembly has been installed. If the gage pressure is less than 700 psi,

it will be necessary to overhaul the pump as outlined in paragraph 75 or 79.

PUMP

Two different types of power steering pumps have been used on All-Purpose tractors; refer to paragraph 75 or 79 depending on whether a hollow sleeve plunger or a roller vane pump is used. For power steering pump on LCG, Utility Industrial or Grove models, refer to paragraph 79.

CONTROL VALVE AND STEERING GEAR ASSEMBLY

39. **ADJUSTMENT.** Before attempting to adjust the steering gear, first make certain that the gear housing is properly filled with lubricant, then disconnect the drag links from pitman arms to remove load from the gear unit. As shown in Fig. FO31, the wormshaft is carried in needle type bearings (14) which require no adjustment. The steering valve thrust bearings (8 & 11) are adjusted by tightening the bearing nut (6), but this should be done only when the valve unit is being serviced. To adjust the sector gear backlash, proceed as follows:

Loosen both sector shaft adjusting screws at least two turns and turn the steering wheel to the mid (or straight ahead) position. Using a screwdriver, turn the adjusting screw for the front sector shaft until a spring scale pull of 1¼-1¾ pounds (measured at rim of steering wheel)

is required to turn the wheel through the mid or straight ahead position. Tighten the adjusting screw lock nut.

With the front sector adjusted, turn the adjusting screw for the rear sector shaft until a spring scale pull of 2-2½ pounds is required when measured as before. Tighten the lock nut.

40. **REMOVE AND REINSTALL.** Remove steering wheel nut and using a suitable puller, remove steering wheel and the felt packing from top of steering column. Disconnect throttle rod link from throttle arm, throttle lever from throttle lever bracket and throttle arm from throttle lever; then, remove throttle lever. On "Select-O-Speed" models, remove PTO control and gear selector assemblies as outlined in paragraph 234. Remove hood and disconnect Proof-Meter cable at instrument panel. Disconnect battery ground cable from battery and battery carrier from steering gear housing. Disconnect ammeter wire from junction block, then disconnect junction block and generator regulator from mounting pad on steering column. Unclip temperature gage and oil pressure gage lines from rear of fuel tank, then disconnect instrument panel from lower rear hood panel. Slide instrument panel over top of steering column and set it on top of fuel tank. Remove headlight switch (if so equipped) from lower rear hood panel, then remove hood rear lower panel. Unclip tail light wire from steering housing and disconnect hydraulic connection from power steering control lines. Cap all open lines and ports to prevent entry of foreign material into power steering system.

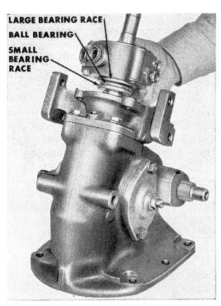

LARGE BEARING RACE
BALL BEARING
SMALL BEARING RACE

Fig. FO29 — When installing upper and lower thrust bearings make sure large race is toward control valve as shown.

SMALL BEARING RACE

BEARING NUT
PRELOAD SPRING
BALL BEARING
LARGE BEARING RACE

Fig. FO30 — View showing method of staking the bearing nut. Be sure to buck-up steering shaft during this operation.

Remove drag links from Pitman arms then remove cap screws retaining steering housing to transmission housing. Pry steering gear housing from dowel pins in transmission and lift steering gear assembly from tractor.

NOTE: On "Select-O-Speed" transmission models, a gasket is used between the steering gear and transmission housings. Gasket should be left in place or opening in transmission covered to keep foreign material from entering transmission and causing serious damage.

Reinstall by reversing the removal procedure. On "Select-O-Speed" models, be sure gasket between steering gear and transmission housings is in good condition. Align steering gear housing on dowel pins, install retaining cap screws and tighten to a torque of 60-70 Ft.-Lbs. Before installing hood side panels, fill and bleed power steering system as outlined in paragraph 37 and at this time, observe all connections for leakage.

41. OVERHAUL CONTROL VALVE.
With the control valve and steering gear assembly removed as outlined in paragraph 40, proceed as follows: Place a scribe line on valve cover (steering column), valve body, valve adapter and steering gear housing to insure proper assembly. Remove cap screws retaining valve cover (steering column) to control valve body and remove valve cover. Unstake the bearing lock nut, then remove and discard the lock nut. Remove the valve spool preload spring. Remove the small bearing race, bearing and the large bearing race and keep these parts together as an assembly so they can be reinstalled in the same position. Lift off valve body and valve as a unit, being careful not to lose plungers, springs or check valves. Remove the lower large bearing race, bearing and small bearing race and keep these three parts together as an assembly.

42. Remove the six centering plungers and three springs from their bores. Remove the two reaction plungers from their bore and note their position so they can be reinstalled in their original location. Be careful not to lose the steel balls. Note which end of the valve (spool) has the groove on the inside diameter and remove the valve. Remove the anti-cavitation valve assembly from the pump return line in valve body. Refer to Fig. FO28.

Inspect brass seats in ports of valve body and renew as needed. Pay particular attention to valve (spool) and valve body for evidence of scoring, galling or other damage. If these parts need renewal, the valve (spool) and valve body must both be renewed as they are not available separately. In-spect all springs for damage or distortion and all plungers for nicks or scratches. Inspect thrust bearings and bearing races for freedom of movement and signs of damage. Renew all parts showing undue wear or damage.

43. To reassemble the control valve proceed as follows: Coat all plungers and control valve (spool) with Lubriplate or equivalent and install the valve (spool) in the valve body with the groove on the inside diameter of valve (spool) at the same end of the valve body as was noted upon removal. Do not use force during this operation as valve (spool) will drop into place when properly aligned. Install anti-cavitation valve assembly in return port of control valve housing. Install the two reaction plungers with balls toward center of bore and in the same location from which they were removed. Install a centering spring with a plunger at each end in the remaining three bores. Now position the lower bearing assembly against the bottom of the control valve assembly in the same order that it was removed; that is, with the large bearing race on the upper side as shown in Fig. FO29. NOTE: The embossed markings at the control valve ports should be on the top side. With the two pump ports positioned just to the right of center, lower bearing assembly and control valve assembly into position on the adapter. Be careful not to lose any of the centering plungers or reaction plungers from control valve body. Install top bearing assembly (with large bearing race on bottom side) on top of control valve. Install preload spring with convex side up and install a new lock nut. Insert the three cap screws through the control valve body and screw them loosely into adapter to prevent control valve from turning; then, temporarily install steering wheel and while gripping same, tighten lock nut until it seats firmly against valve (spool). Then back-off nut about ⅛-turn and while bucking up steering shaft, stake nut in this position as shown in Fig. FO30. Before installing valve cover (steering column) check shaft seal. If a new seal is used, coat with Lubriplate and install with lip facing control valve. Renew "O" ring in casting face and install valve cover with generator regulator mounting pad forward. Torque cap screws to 20 Ft.-Lbs.

Reinstall complete unit on tractor as outlined in paragraph 40. Fill and bleed system as in paragraph 37.

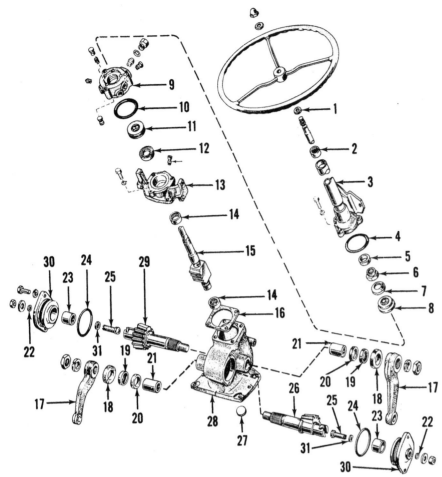

Fig. FO31 — Exploded view of the power steering gear and control valve assemblies which are used on All-Purpose, Utility Industrial and LCG models. Power steering gear used on Grove type is similar. Steering gear portion of this assembly is very similar to those used on models wihout power steering.

1. Seal	11. Lower thrust bearing	21. Bushing
2. Bearing	12. Oil seal	22. Lash adjuster packing
3. Gear housing cover	13. Adapter	23. Bushing
4. "O" ring	14. Needle bearing	24. "O" ring
5. Oil seal	15. Steering shaft	25. Lash adjuster
6. Bearing lock nut	16. Gasket	26. Sector shaft, single
7. Preload spring	17. Pitman arm	27. Expansion plug
8. Upper thrust bearing	18. Dust seal	28. Housing
9. Control valve assembly	19. Packing retainer	29. Sector shaft, double
10. "O" ring	20. Packing	30. Side cover
		31. Shim

NOTE: Late production power steering models are equipped with heavy duty steering sector gears (26 & 29—Fig. FO31). Only the heavy duty sector gears will be available for servicing all models when current stocks of standard sector gears are exhausted. If necessary to renew a standard sector gear using a heavy duty gear, both sector gears must be renewed. Gears may be identified as follows: Standard sector gear (26) has 7 full teeth; heavy duty sector gear (26) has 5 full teeth. Standard sector gear (29) has 6 full teeth and 3 rack teeth; heavy duty sector gear (29) has 4 full teeth and 3 rack teeth. Service procedures remain unchanged.

44. OVERHAUL STEERING GEAR.
With steering gear assembly removed as in paragraph 40 and the control valve removed as in paragraph 41, proceed as follows: Drain oil from steering gear housing and clamp assembly in vise. Use a suitable puller and remove pitman arms and felt seals. Remove right sector shaft side cover and remove sector shaft from side cover by turning the adjusting screw clockwise. Remove left side sector shaft cover in the same manner. Remove adapter, then pull steering shaft and ball nut assembly from housing.

Inspect needle bearings in bottom of steering gear housing and adapter. Bearings should be smooth, highly polished and free to rotate in their retainers. If new bearings are installed, press on trade-marked side of bearing and press bearing in until outer edge is just below chamfer on face of casting. DO NOT press either bearing in so far that it bottoms. In-

spect seal in adapter and if necessary, remove with an offset screwdriver. Coat new seal with Lubriplate and install with lip facing control valve. Remove seals (packing) from sector shaft bores in steering gear housing and inspect bushings. New bushings have an inside diameter of 1.1245-1.1250 and if carefully installed, should need no final sizing. Inspect bushings in sector shaft side covers for excessive wear. If necessary to renew any of the bushings, it is recommended that they be removed with a suitable puller and installed with a press. New side cover bushings have an inside diameter of 1.1255-1.1260 and should require no final sizing if carefully installed. Inspect sector shafts and sector gear teeth for signs of wear or chipping and renew as necessary. Inspect steering shaft and ball nut for nicks, scratches or signs of wear on the ball nut rack teeth. If damaged, steering shaft and ball nut must be renewed as an assembly.

When reassembling, use all new "O" rings and gaskets and proceed as follows: Place a shim (31—Fig. FO31) over threaded end of each sector shaft adjusting screw (25) to obtain a clearance of 0.000-0.002 between head of adjusting screw and bottom of slot in sector shafts. Shims are available in thicknesses of 0.063, 0.065, 0.067 and 0.069. Now pull sector shafts into their respective covers by using the adjusting screws as jack screws.

Center ball nut on steering shaft and install steering shaft in housing. Ball nut is properly centered when grooves in steering shaft extend equally beyond each side of ball nut. Install left sector shaft and cover first in the following manner: With ball nut centered and the block tooth of sector shaft in upper-most position, mesh the middle tooth of sector shaft with the middle tooth of ball nut rack. Install cap screws and torque to 25-30 Ft.-Lbs. Install the right sector shaft and side cover so that the fourth tooth of right sector shaft meshes with the fourth groove of the left sector shaft and tighten the side cover cap screws to 25-30 Ft.-Lbs. Use new gasket and install adapter with oil filler hole forward and tighten cap screws to 20-25 Ft.-Lbs. Position felt seals over ends of sector shafts and install Pitman arms.

Install control valve assembly and cover (steering column) as outlined in paragraph 43, adjust the unit as in paragraph 39 and install entire steering gear assembly on tractor by reversing procedure given in paragraph 40. Fill steering gear housing with specified lubricant and bleed power steering system as outlined in paragraph 37.

Fig. FO32 — View of All-Purpose type power cylinder. Cylinders are right and left hand assemblies and must be identified prior to removal.

POWER CYLINDERS

All-Purpose and LCG Models

45. **REMOVE AND REINSTALL.** Power cylinders are right and left hand assemblies and must be identified prior to removal. With the power cylinders properly identified, place a drain pan under cylinder to be removed and disconnect the hydraulic lines. Remove nuts from both front and rear end assemblies and using a soft faced hammer, bump same from their mountings. Cylinder can be cleared of oil by moving piston rod back and forth and allowing oil to drain from ports.

46. **OVERHAUL.** Loosen clamp and unscrew forward end assembly from piston rod. Remove snap ring (FO 32) and extend piston rod to the end of its travel; then, remove the metal scraper washer, leather wiper washer and the aluminum bushing. Seal will normally come out by actuating the piston rod; however, if it does not, it can be removed with an awl or some other sharp pointed tool.

Clean cylinder and rod assembly and inspect for leakage at seams, gouging or distortion of piston rod and/or damage of threads at port end of cylinder. If any damage is found the entire cylinder assembly must be renewed as only seals and rod end assemblies are catalogued for service.

Reassemble by reversing the disassembly procedure and cover threads on end of piston rod with masking tape prior to installing seal. Seal and leather wiper should be soaked in Ford M2C41 hydraulic oil before installation. Raised center portion of metal scraper ring must face outward. Power cylinders are installed with ports of cylinder downward. After installing hydraulic lines, fill and bleed system as outlined in paragraph 37.

Utility Industrial and Grove Types

47. **REMOVE AND REINSTALL.** Procedure for removing and reinstalling power steering cylinders on Utility Industrial and Grove types is same as outlined for All-Purpose models in paragraph 45.

48. **OVERHAUL.** Loosen clamp and unscrew front end assembly from cylinder rod. Remove snap ring (2— Fig. FO33 or FO34) and extend cylinder rod to end of travel. The seal unit will usually come out as the rod is extended. If not, it may be pried out with an awl or other sharp pointed tool.

Early production cylinders were equipped with an aluminum seal retainer (8—Fig. FO33) using two "O" rings (6 and 7), a back-up ring (5) and a wiper ring (4). Originally, the seal service kit consisted of all these parts packed loosely in a carton. When using these kits, be sure the internal "O" ring, back-up ring and wiper are installed in proper location as shown in Fig. FO33. Later service kits have the internal "O" ring, back-up ring and wiper already installed in the seal retainer; with these kits, it is only necessary to install the external "O" ring (7) on the retainer, lubricate the assembly and slide it into place. Be sure the snap ring (2) is securely installed with sharp side of ring to outside.

Late production cylinders are equipped with three rubber sealing rings as shown in Fig. FO34. (Early and late type cylinder assemblies are interchangeable as complete units; however, the seal kits are not inter-

changeable). Be sure the three rubber sealing rings are installed in order shown in Fig. FO34. The rings are available in a kit only and are not serviced separately.

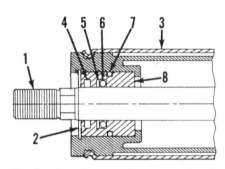

Fig. FO33 — Cross-sectional view of early Utility Industrial power steering cylinder showing type of seals used on cylinder rod end.

1. Cylinder rod	5. Back-up ring
2. Snap ring	6. "O" ring
3. Cylinder assembly	7. "O" ring
4. Wiper	8. Seal retainer

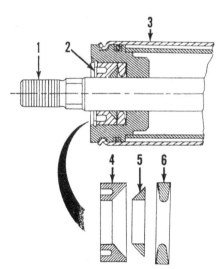

Fig. FO34 — Cross-sectional view of late production Utility Industrial power steering cylinder showing seal arrangement.

1. Cylinder rod	4. Outer seal ring
2. Snap ring	5. Center seal ring
3. Cylinder assembly	6. Inner seal ring

POWER STEERING SYSTEM
(Early Heavy Duty Industrial With Full Power Steering)

Early Heavy Duty Industrial Tractors incorporate a "Full Power" steering system. No mechanical linkage exists between the activating wheel and the tractor front wheels, the entire steering action being controlled hydraulically. The system consists of a pump and reservoir assembly, power cylinder and control valve assembly, an actuating cable and an actuating shaft, housing and wheel assembly. See Fig. 35 for a detailed view of the power steering components. If tractor has been converted to drag link (power assist) steering, refer to paragraph 60.

NOTE: The maintenance of absolute cleanliness of all parts is of utmost importance in the operation and servicing of the hydraulic power steering system. Of equal importance is the avoidance of nicks and burrs on any of the working parts.

LUBRICATION AND BLEEDING

49. The power steering system is a self-bleeding unit. Whenever service has been performed on any of the component parts, reconnect and tighten all hydraulic lines, fill the pump reservoir to within ½-inch of the top with Ford Hydraulic Fluid, M2C41, and cycle the system several times at idle engine speed to bleed air from the cylinder and lines. Shut off the engine and refill the reservoir to within ½-inch of the top.

TROUBLE-SHOOTING

50. The following are common troubles which may be encountered in the operation of the power steering system. The procedure for correcting most of the troubles is evident; for those not readily remedied, refer to the appropriate subsequent paragraphs.

LOSS OF TURNING ACTION. May be due to inoperative pump or control valve. Look for; broken drive belt, low fluid level, damaged pump, sticking flow control or relief valve, sticking control valve or broken actuator cable.

UNEQUAL TURNING RADIUS. Usually due to improperly adjusted actuating cable or tie rod. Could also be caused by damaged steering cylinder.

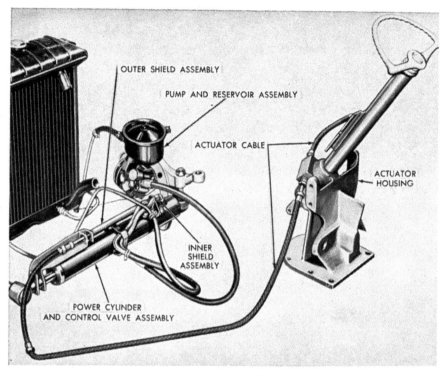

Fig. FO35—Full power steering assembly used on Heavy Duty Industrial tractors. Actuator cable moves the control valve located in the power steering cylinder whenever steering wheel is moved. System is neutralized by cable stops at each end of stroke so that relief valve is not activated.

Fig. FO36 — Method of teeing pressure gage and shut-off valve into pump pressure line to test pressure of Heavy Duty power steering pump with full power steering. Because of design of system, relief valve cannot be activated in normal operation.

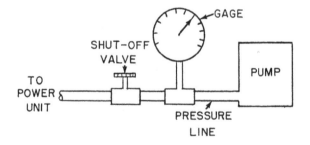

ERRATIC STEERING CONTROL. May be due to low fluid level or wrong fluid in system. Could also be caused by faulty or sticking flow control or relief valve, loose drive belt, plugged filter or worn or damaged pump.

WHEELS SHIMMY. Usually caused by foaming fluid in system due to low fluid level, wrong fluid in system, plugged filter or worn shaft seal. Could also be caused by loose drive belt, faulty control valve or loose connection between actuator cable and operating stud.

SYSTEM OPERATING PRESSURE

51. A presure test of the hydraulic circuit will disclose whether the pump or some other unit in the system is malfunctioning. To make such a test, proceed as follows: Connect a pressure test gage and a shut-off valve in series with the pump pressure line as shown in Fig. FO36. Note that shut-off valve is connected in the circuit between the gage and the power cylinder. Open the shut-off valve and run the engine at idle speed until oil is warmed to operating temperature. With engine running at 1300 rpm,

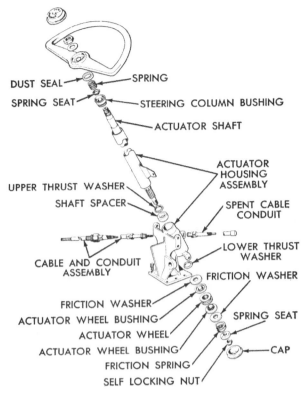

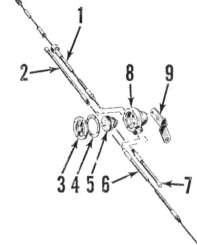

1. Lower cable & conduit
2. Conduit extension
3. Housing cover
4. Gasket
5. Reduction wheel
6. Upper cable & conduit
7. Conduit extension
8. Housing
9. Bracket

Fig. FO39 — Exploded view of actuator cable, reduction unit and associated parts.

Fig. FO37 — Exploded view of actuator assembly used on Heavy Duty Industrial full power steering. See text for details.

Fig. FO38 — Assembled view of 2:1 reduction un.t used in actuator cable.

slowly close the shut-off valve and retain in closed position only long enough to observe the gage reading. Pump may be seriously damaged if valve is left closed longer than 10 seconds. If gage reading is 1250 psi or more, with the valve closed, the pump and relief valve are satisfactory. Factory setting of relief valve is 1550 psi, and is not adjustable.

ACTUATOR

52. **OPERATION.** The actuator shaft and wheel assembly is mounted on the tractor in place of the conventional steering gear. The control valve, located in the power cylinder assembly is operated by the wheel through the actuator cable assembly. Movement of the steering wheel in either direction activates the control valve through a 2:1 reduction unit mounted on left rear of cylinder block.

The rod end of the power cylinder assembly is anchored to tractor axle, and resultant movement of the power cylinder will neutralize the control valve and maintain position until further movement of the steering wheel again actuates the valve. Fixed limit

stops, on the cylinder end of the actuator cable and shield, limit the movement of the cable at each end of the cylinder stroke. If all adjustments of the actuator cable are correct the control valve will always be neutralized before the steering arms contact the axle stops. Because of this arrangement, the safety valve will not come into operation because of excessive cramping of the front wheels. The cable actuator wheel on the lower end of the steering shaft is fixed to the steering shaft by two friction washers. Further turning of the steering wheel after full right or left turn, or when the tractor engine is stopped will overcome the friction drag of the washers and the steering wheel will turn without further loading of the actuator cable.

53. **ADJUSTMENT.** Friction drag of the actuator wheel on the steering shaft is correct when 25-30 pounds

effort is required to turn the steering wheel when the engine is not running.

CAUTION: Cable or associated parts may be damaged if substantially more pressure is applied. If steering wheel cannot be turned with reasonable effort, readjust as follows:

Remove the battery and loosen battery carrier. Pry the dust cap from lower end of actuating housing. Loosen cable conduit and disengage cable from actuator wheel by turning steering wheel counter-clockwise. Attach a pull scale to one end of a short piece of spare cable. Feed the spare cable into actuator wheel. Adjust the self locking nut at lower end of steering wheel shaft until actuator wheel will turn on shaft when a pull of 75 pounds is exerted on actuator cable. Early production half-circle steering wheel can be repositioned to the operator's satisfaction as follows: Start the tractor engine and position the front wheels straight ahead. Stop the engine and reposition the steering wheel as desired by overcoming the friction drag.

54. R&R AND OVERHAUL. The steering shaft and actuator components may be removed from the actuator housing without removing housing from the tractor. If necessary to remove actuator housing for other reason, follow general procedures outlined in paragraph 63 after disconnecting actuator cable as outlined in paragraph 56. To remove steering shaft and actuator components with housing in place, proceed as follows: Remove the battery and loosen battery carrier. Using Fig. FO37 as a guide for parts identification, proceed as follows: Disconnect the actuator cable conduit from actuator housing and remove the cable by rotating steering wheel counter-clockwise until cable is free of the actuator wheel. Remove dust cap from lower end of housing, then remove the self-locking nut by holding it with a wrench while turning the steering wheel counter-clockwise. Remove lower spring seat and extract the remaining actuator wheel components from lower end of housing.

Pry the hub cover from the steering wheel and remove the nut, washer and steering wheel. Remove the spring, dust seal and spring seat from steering column and lift the actuator shaft out of housing. If necessary, tap bottom of shaft with a soft faced hammer. The upper bearing will be removed with the steering shaft.

Clean all parts in a suitable solvent and inspect for wear or damage.

Drive steering shaft spacer out the bottom of the actuator housing if renewal is indicated.

To reassemble, reverse the disassembly procedure. Install the steering shaft spacer with chamfer end up and, using a puller screw and washer, draw it into the housing bore until bottom face of spacer is flush with its bore in the housing. Coat bearing surfaces and lower spline of steering shaft and pack lower end of the actuator housing with (Ford) M-1C-18 low temperature grease or its equivalent. Set pre-load on friction spring as outlined in paragraph 53. Adjust actuator shaft bearings by means of the nut at steering wheel end of shaft. Bearing pre-load is correct when a rolling torque of 7 inch-pounds is obtained without the cable installed.

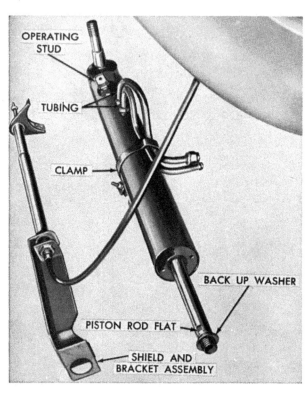

Fig. FO40 — Cylinder and valve assembly disconnected from tractor and actuator cable prior to servicing.

Fig. FO41 — Using pin type spanner wrench to remove bearing and seal assembly from power steering cylinder.

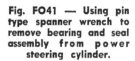

55. ACTUATOR CABLE AND REDUCTION GEAR. Refer to Fig. FO-38. The control cable incorporates a 2:1 reduction unit which attaches to engine block mounting bolts as shown. The complete cable and reduction unit should be removed for service as follows:

56. REMOVAL. Remove the boot from operating stud on power cylinder. Disconnect lower actuator cable conduit from cylinder slide and remove the two nuts which secure lower cable to valve operating stud.

Disconnect upper actuator cable conduit from actuator housing and disengage cable from steering shaft by turning steering wheel counter-clockwise.

Disconnect both cable conduits from tractor frame; then unbolt and remove the complete reduction unit with cables attached.

57. OVERHAUL AND ADJUST. Remove the conduit extensions (2 and 7—Fig. FO39) and pull the cables from upper and lower conduits (1 and 6). Remove cable conduits from reduction housing (8). Remove cover (3) and wheel (5) from housing. Inspect the parts and renew any which are worn or damaged.

When reassembling, coat the reduction wheel (5) liberally with Lubriplate and reassemble the reduction unit. Coat both actuator cables with Lubriplate and install in their respective conduits. Attach lower conduit to cylinder and attach lower end of cable to valve operating stud. Adjust the two nuts so that ¼-inch of cable extends through operating stud. Tighten the nuts; then tighten and stake the the conduit lower retaining nut.

Insert upper end of lower cable in inner, front opening of reduction housing. Push housing on cable, allowing reduction wheel to turn, until housing contacts lower cable conduit, then secure conduit to reduction housing. Mount the reduction gear assembly in place on tractor and attach lower conduit to side rail bracket.

Start the tractor engine and grasp the free end of lower cable where it extends through reduction housing. Pull cable through housing until front wheels are turned fully to the right. Attach upper cable conduit to outer, rear opening of reduction housing; then thread upper cable through reduction gear until approximately ⅜-inch of free end of cable extends through housing. With engine running and wheels at full right turn, ⅛—⅜-inch of upper cable should extend through housing while tension is applied to upper cable. Adjust by threading upper cable in or out of housing.

Attach upper cable to actuator housing, turning steering wheel clockwise until the conduit contacts actuator housing. Complete the assembly by reversing the disassembly procedure.

After assembly is complete, start the engine and check to see that front wheels turn fully to right and left an equal distance; and that right steering arm does not contact the turning stops. Adjust by repositioning the control valve operating stud on lower cable.

PUMP AND RESERVOIR

Refer to paragraph 79 for information on servicing the roller vane type power steering pump.

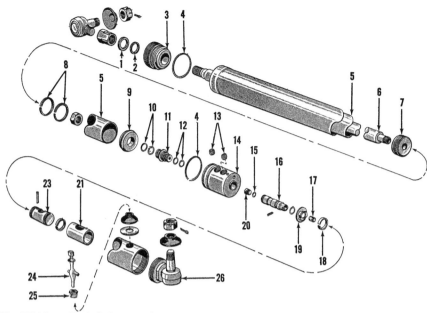

Fig. FO42 — Exploded view of early production control valve and power cylinder assembly used on Heavy Duty Industrial tractor full power steering system.

1. Outer seal	8. Piston ring	14. Valve housing	20. Plug
2. Inner seal	9. Retainer	15. "O" ring	21. Guide spacer
3. Shaft bearing	10. "O" rings	16. Valve spool	23. Valve guide
4. "O" ring	11. Inner plug	17. Connecting pin	24. Operating stud
5. Inner tube	12. "O" rings	18. Bushing	25. Bushing
6. Cylinder rod	13. Seat	19. Retainer	26. Cylinder end
7. Piston			

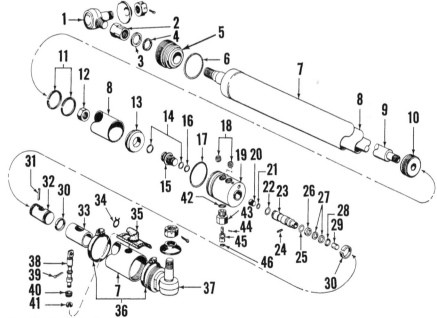

Fig. FO43 — Exploded view of latest type full power steering system cylinder used on Heavy Duty Industrial tractors. Note by-pass valve (45) and related parts. Leather sealing boot (35) can be installed on early cylinders using clamps (34 and 36).

1. End assembly	13. Retainer	25. "O" ring	36. Clamps
2. Sleeve	14. "O" rings	26. Key ring	37. Cylinder end
3. Outer seal	15. Plug	27. Washers	38. Actuating stud
4. Inner seal	16. "O" ring	28. Retainer	39. Pivot pin
5. Shaft bearing	17. "O" ring	29. Connecting pin	40. Grommet
6. "O" ring	18. Tubing seats	30. Nylon bushings	41. Snap ring
7. Outer tube	19. Valve housing	31. Pin	42. "O" ring
8. Inner tube	20. Plug	32. Stud support	43. Valve housing
9. Cylinder rod	21. "O" ring	33. Support guide	44. Pin
10. Piston	22. "O" ring	34. Spring clamp	45. By-pass valve
11. Piston rings	23. Valve spool	35. Leather boot	46. "O" ring
12. Piston nut	24. Pin		

CONTROL VALVE AND
POWER CYLINDER ASSEMBLY

NOTE: The outlined procedures refer specifically to early type unit shown in Fig. FO42. Except for obvious differences, procedures for late type unit (Fig. FO43) are similar.

58. R&R AND OVERHAUL. Disconnect tie rod end from left steering arm and swing tie rod back to gain access to the cylinder rod end connection. Place a drain pan under the pressure line connections, disconnect pressure and return lines at the cylinder tubes and allow pump and cylinder to drain into the pan. While the system is draining, disconnect the control cable from control valve operating stud and cable shield from cylinder housing. Disconnect cylinder end of the power steering cylinder from right steering arm by removing the cotter pin and castellated nut. Lower right end of cylinder to the floor. Fit a suitable end wrench over the machined flat at outer end of the piston rod and thread the rod out of the rod end attached to front axle. Slip actuating cable shield and bracket assembly off the end of the piston rod and place the cylinder on a clean bench for disassembling. See Fig. FO40.

59. Clamp the cylinder assembly in a vise as shown in Fig. FO41, but do not clamp the vise at the control valve ports. Use only enough pressure to hold the cylinder, being careful not to distort the outer tube. Using a suitable spanner wrench, unscrew the shaft bearing (3—Fig. FO42) from the outer tube. The bearing (3), inner tube (5), piston (7) and rod (6) can then be withdrawn from the outer tube. If inner tube does not come with the rod, remove it after the rod has been withdrawn. Remove the inner tube from the rod and bearing assembly, carefully remove all paint, burrs and foreign material from outer end of piston rod and slide the bearing off the end of the rod. Remove tubing and clamp from cylinder assembly and remove the operating stud (24) by pulling it upward out of the cylinder. Unscrew the cylinder end assembly (26) from outer tube, remove tube from vise and tap rod end of cylinder on the bench to jar out the valve body and operating mechanism. Remove the guide spacer (21) from valve guide and drive out the roll pin securing guide to valve spool.

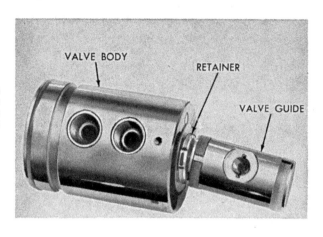

Fig. FO44 — Bump valve body, spool and retainer from outer cylinder as a unit after removing piston and inner tube.

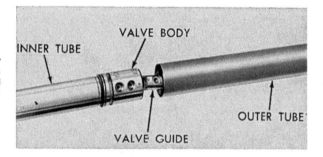

Fig. FO45 — Install valve body, guide, inner tube, piston, shaft and bearing in outer tube as a unit.

Clean the parts with a suitable solvent and discard all "O"-rings and seals. Renew any other parts which are worn, scored, or otherwise damaged.

Lubricate all parts in clean oil before reassembly. Install the brass connecting pin to valve spool with a roll pin and install plug in opposite end of spool. The plug will position itself during operation, so proper positioning at this time is not important. Start the valve spool in the valve body from cylinder end of body and push spool through only far enough to clear pin hole in yoke end of the spool, to prevent damage to "O"-ring on forward end of spool. Reinstall the inner plug in valve body, making sure it is fully seated. Place the "O"-ring and retainer over the yoke end of spool and reinstall the guide, securing same to spool with the roll pin.

Reassemble the piston rod and install bearing over the rod. Clamp outer end of piston rod in a vise with the piston end up. Make sure to clamp the rod in the vise at the milled flats.

Assemble the inner tube in position over the piston and secure outer end in the bearing. Connect valve body assembly to the inner tube by the inner plug and insert rod and control valve assembly into the outer cylinder. Screw the bearing into the outer cylinder a few turns and check to see that ports are in line with holes in the outer cylinder. Install the spacer tube over the valve guide and align and insert the operating stud. Screw the cylinder end assembly into the control valve end of outer tube until the peening mark made on the threads at original assembly align with the 1/8 inch hole in the outer tube. Then, turn the cylinder rod bearing into the outer tube and tighten to a torque of 200-230 Ft. Lbs.

Stake both ends of the cylinder to the outer tube and reinstall unit on tractor by reversing the removal procedure. Refill reservoir with Ford Hydraulic Fluid, M2C41; then bleed system as outlined in paragraph 49, and adjust as in 53.

POWER STEERING SYSTEM

(Heavy Duty Industrial With Drag Link, Power-Assist, Steering)

Late Heavy Duty Industrial models (and modified early models) incorporate a "Power-Assist" steering system. Mechanical linkage is used between the steering gear and front wheels. A control valve assembly is an integral part of the steering gear unit (refer to Fig. FO46). The power assist cylinder is mounted at rear side of front axle and is connected to the front axle and steering arm on right front wheel spindle; refer to Fig. FO5.

NOTE: The maintenance of absolute cleanliness of all parts is of utmost importance in the operation and servicing of the hydraulic power steering system. Of equal importance is the avoidance of nicks or burrs on any of the working parts.

LUBRICATION AND BLEEDING

60. Whenever service has been performed on any of the component parts, proceed as follows to bleed all air from system: Fill reservoir to top with Ford M-2C-41 Hydraulic Fluid. Detach piston rod end of power steering cylinder from front axle. With engine running at low RPM, keep filling reservoir while turning the steering wheel from stop to stop. A minimum of six cycles from stop to stop is usually required to purge all air from the system. Reattach cylinder piston rod to front axle. Turn steering wheel to full right position. Fluid level should then be approximately 5/8-inch from top of reservoir. After correcting the fluid level, securely install reservoir cover.

SYSTEM OPERATING PRESSURE

61. A pressure test of the power steering system will disclose if the pump is developing sufficient pressure to be considered in satisfactory condition. To make such a test, proceed as follows:

Connect a pressure test gage in series with the pump pressure line. Run the engine at low idle speed until the power steering fluid is warmed to normal operating temperature. Advance the engine speed to high idle RPM and turn the steering wheel against stop in either direction and hold in this position only long enough to observe the gage reading. NOTE: Pump may be seriously damaged if

steering wheel is held in this position for an extended period of time. Maximum pressure gage reading should be 1350 psi. If reading is near this pressure, pump and relief valve can be considered OK.

If gage reading is 700-800 psi, or about 1550 psi, this would indicate that incorrect relief valve part number has been installed in pump. If gage reading is abnormally high, incorrect installation of relief valve would be indicated. Refer to paragraph 80 for further information.

If gage reading is low, and relief valve malfunction has been eliminated as cause, it will be necessary to remove and overhaul the pump as outlined in paragraphs 79 and 80.

PUMP AND RESERVOIR

Refer to paragraph 79 for information on the roller vane type power steering pump.

CONTROL VALVE AND STEERING GEAR ASSEMBLY

62. **ADJUSTMENT.** To adjust sector shaft end play, disconnect the drag link from pitman arm on steering gear to remove any load from the gear unit. Turn the steering wheel to the mid (straight ahead) position and check end play of sector shaft. If end play is perceptible, or if steering gear appears to bind as steering wheel is turned, readjust the sector shaft end play as follows: With steering wheel in mid position, loosen the locknut (30—Fig. FO46) and adjust screw (25) so that there is no perceptible end play of sector shaft, and no binding condition occurs as the steering wheel is turned past mid (straight ahead) position.

End play of the steering (worm) shaft is not adjustable as the shaft rides in straight needle roller bearings (14 and 20). End float of the steering shaft is necessary to actuate the power steering control valve spool (5).

Reattach the drag link to pitman arm on steering gear and check length and position of drag link as outlined in paragraph 17.

63. **REMOVE AND REINSTALL.** Remove steering wheel. Disconnect throttle rod link from throttle arm, throttle lever from throttle lever bracket and throttle arm from throttle lever; then, remove throttle lever. On "Select-O-Speed" models, remove PTO control and gear selector assemblies as outlined in paragraph 234. Remove hood and disconnect Proof-Meter cable at instrument panel. Disconnect battery ground cable from battery and battery carrier from steering gear housing. Disconnect ammeter wire from junction block, then disconnect junction block and generator regulator from mounting pad on steering column. Unclip temperature gage and oil pressure gage lines from rear of fuel tank, then disconnect instrument panel from lower rear hood panel. Slide instrument panel over top of steering column and set it on top of fuel tank. Remove headlight switch from lower rear hood panel, then remove hood rear lower panel. Unclip tail light wire from steering housing and disconnect hydraulic lines from power steering control valve. Cap all open lines and ports to prevent entry of foreign material into power steering system. Remove drag link from pitman arm and remove cap screws retaining steering housing to transmission housing. Pry steering housing loose from dowel pins in transmission housing and lift steering gear assembly from tractor. NOTE: On "Select-O-Speed" transmission models, a gasket is used between the steering gear and transmission housings. Gasket should be left in place or opening in transmission covered to keep foreign material from entering the transmission and causing serious damage.

Reinstall by reversing the removal procedure. On "Select-O-Speed" transmission models, be sure the gasket between transmission and steering gear housing is in good condition. Align steering gear housing on dowel pins, install retaining cap screws and tighten to a torque of 60-70 Ft.-Lbs. Fill and bleed power steering system as outlined in paragraph 60 before installing hood side panels, so that hydraulic connections at power steering control valve can be observed for leakage.

64. OVERHAUL CONTROL VALVE.
Refer to paragraph 41 for information on overhauling the control valve. Except that the pump return port anti-cavitation valve shown in Fig. FO28 is not used on Heavy Duty Industrial power steering, the power steering control valve is similar to that used on All-Purpose type tractors. Through-bolt banjo type fittings are used to connect the hydraulic lines to the control valve; tubing flare fittings are used on All-Purpose tractors.

65. OVERHAUL STEERING GEAR.
With the steering gear assembly removed as outlined in paragraph 63 and the control valve removed as outlined in paragraph 41, proceed as follows: Drain oil from steering gear housing and use suitable puller to remove pitman arm from sector shaft. Unbolt gear housing (31—Fig. FO46) from bushing housing (34). Unbolt end cover (29) from gear housing and withdraw cover, adjusting screw (25) and sector shaft (24) unit. Unbolt adapter (13) from gear housing and withdraw steering (worm) shaft (16) and ball nut (19) assembly.

Disassemble end cover and sector shaft unit by removing locknut (30) and turning adjusting screw inward until threads are disengaged from end cover. Bushing (27) in end cover is presized and if carefully installed, reaming should not be necessary. Inside diameter of bushing (new) is 1.1255-1.1260. Inspect sector shaft for signs of wear, chipped gear teeth or other defect and renew if necessary. Shim (26) on adjusting screw (25) is available in thicknesses of 0.063, 0.065, 0.067 and 0.069. When reassembling, use shim that will provide zero to 0.002 clearance between head of adjusting screw and slot in end of sector shaft.

Inspect bushing (32) and oil seal (33) in gear housing and renew if necessary. Bushing is pre-sized and should not require reaming if carefully installed. Install oil seal (33) with lip of seal towards inside of housing. Inspect needle bearing (20) in bottom of gear housing. If worn or damaged, renew as follows: After removing old bearing, press new bearing into bore with lettered side up (towards bearing driver). Do not press bearing in so far that it contacts shoulder in bottom of housing.

Inspect needle bearing (14) in adapter (13). If bearing is worn or damaged, remove it and install new bearing as follows: Press new bearing into bottom of adapter with let-

tered side of bearing cage down (towards bearing driver). Press bearing in until just below flush with lower face of adapter, but not so far that it contacts spacer washer (12). If seal (11) in adapter is damaged, pry old seal out and carefully press new seal into place with lip towards control valve.

Inspect steering (worm) shaft (16) and ball nut (19) for wear or other damage; if necessary to renew either part, a complete new steering shaft and ball nut assembly must be installed. The 106 steel balls (22), ball guides (18) and ball guide clamp (17) are available for service. To disassemble the unit, remove the clamp and ball guides. Separate the guides and drop the balls out into a pan. Shake the steering shaft and ball nut over the pan to remove the remaining balls. To reassemble, count the steel balls out into two groups of 53 each. Slide the ball nut over the steering worm, align hole in ball nut with groove in worm and drop balls from one group of 53 into the hole. Work the balls into the hole with a

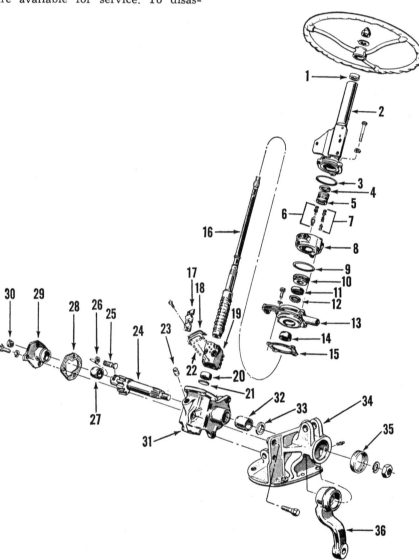

Fig. FO46 — Exploded view of steering gear and power steering control valve assembly used on Heavy Duty Industrial models with drag link (power assist) steering. Not shown is bearing nut (refer to 6—Fig. FO31), spring washer (7—Fig. F031) and upper thrust bearing (same as 10).

1. Bearing	9. "O" ring	19. Ball nut	28. Gasket
2. Steering shaft housing	10. Thrust bearing	20. Bearing assy.	29. End plate
3. "O" ring	11. Oil seal	21. Expansion plug	30. Lock nut
4. Oil seal	12. Washer	22. Recirculating balls	31. Gear housing
5. Control valve spool	13. Valve adapter	(106)	32. Bushing
6. Reaction plungers &	14. Bearing assy.	23. Filler plug	33. Oil seal
anti-cavitation valves	15. Gasket	24. Sector shaft	34. Housing
7. Centering springs &	16. Steering (worm) shaft	25. Adjusting screw	35. Bushing
plungers	17. Clamp	26. Shim	36. Pitman arm
8. Control valve housing	18. Ball return guides (4)	27. Bushing	

small rod until circuit is filled from bottom of hole to bottom of opposing hole. It may help to carefully rotate the steering worm back and forth while filling the hole. Using a light grease such as Lubriplate, stick the remaining balls from the group of 53 into the split ball guide (18) and insert guide in place in filled circuit holes. Install the second group of 53 balls in similar manner, then install the guide clamp (17).

Center ball nut on steering shaft and install shaft in housing. Assemble sector shaft in end cover and insert sector shaft into housing with middle tooth of sector gear engaged in middle tooth groove in ball nut. Install the control valve adapter using a new gasket (15). Renew bushing (35) in housing (34) if necessary and install housing on gear housing (31). Center the steering gear and install pitman arm (36) with arm downward. Fill the gear housing to proper level with SAE 80 gear lubricant.

Install control valve assembly and steering shaft cover as outlined in paragraph 43. Adjust sector shaft end play as outlined in paragraph 62 and reinstall unit on tractor as outlined in paragraph 63.

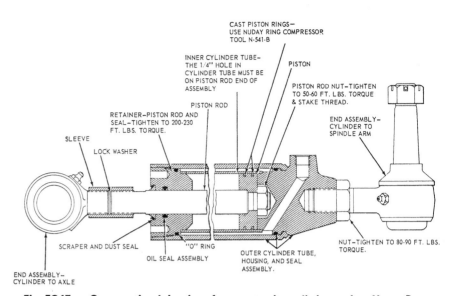

Fig. FO47 — Cross-sectional drawing of power steering cylinder used on Heavy Duty Industrial models with drag link (power assist) steering. Note torque values given for piston rod retainer, piston rod nut and end assembly nut.

POWER CYLINDER

66. REMOVE AND REINSTALL. Place a drain pan under cylinder and remove hydraulic hoses from cylinder head. Cycle the steering gear back and forth to clear cylinder of oil. Remove the cylinder from axle and steering arm attaching points.

After reinstalling cylinder, fill and bleed the power steering system as outlined in paragraph 60. NOTE: Do not attach rod end of cylinder to axle until bleeding procedure has been completed.

67. OVERHAUL CYLINDER. Refer to Fig. FO47 for cross-sectional drawing of the power steering cylinder. The drawing points out parts of the cylinder which are serviced and lists special torque specifications.

All service to the cylinder assembly is performed from the piston rod end. Clamp the hose port body in a soft-jawed vise and remove the rod guide with a spanner wrench as shown in Fig. FO41. A ring compressor, Nuday Tool No. N-541, is required to compress the cast iron piston rings when reassembling unit.

Disassembly and assembly procedure is evident from inspection of unit and reference to Fig. FO47.

POWER STEERING SYSTEM

(Row Crop Type)

The power steering system used on Row Crop type tractors is of the linkage booster type, used in conjunction with the conventional type steering gear mounted on transmission (See Fig. FO17) and in front pedestal (Fig. FO10 or Fig. FO11). Refer to Fig. FO48. The system is so designed that should a hydraulic power assist failure occur, the tractor can be steered with the mechanical linkage.

Basically, the power steering system consists of an engine driven pump incorporating a flow control and relief valve, a reservoir and filter assembly, a double acting cylinder and a control valve.

LUBRICATION AND BLEEDING

68. The power steering system has only one fitting (F—Fig. FO48), located on the lower side of the sleeve and flange assembly. Lubricate this fitting with pressure gun grease after every 10 hours of operation. NOTE: This is a special fitting with short threads. Installation of normal length fitting in this location can damage power steering cylinder and/or lock control sleeve in cylinder.

To fill and bleed the system, turn the front wheels to the extreme left position, remove the reservoir cover or filler cap and fill the reservoir to

within ½-inch of the top or to full mark on dipstick with Ford hydraulic oil M2C41. Start engine and run at low idle speed. (Do not accelerate.) Slowly turn the wheels from one extreme position to the other and observe the fluid for bubbles and turbulence during the turning operation. The presence of bubbles and turbulence indicates that there is air in the system. Continue turning the wheels from one extreme position to the other until all the air has been removed from the system; then turn the wheels to the extreme left and refill the reservoir to within ½-inch of the top on rotor or roller vane type pump or to "full" level indicated on dipstick on hollow plunger type pump. Reinstall the reservoir cover or filler cap.

TROUBLE-SHOOTING

69. The following paragraphs outline possible causes and remedies for troubles in the power steering system.

LOSS OF POWER ASSIST could be caused by:

1. Faulty or improperly adjusted pump drive belt.

2. Insufficient fluid in system or

Fig. FO48 — View showing power steering installation on Row Crop tractors. Measured distances between X and S, and Y and S, must be maintained to obtain proper turning radius.

C. Clamp
F. Grease fitting
S. Ball stud
X. Front drag link end
Y. Rear drag link end

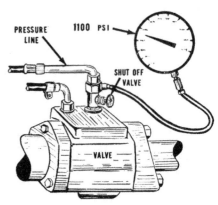

Fig. FO49 — View showing method of installing gage and shut-off valve to check operating pressure of power steering hydraulic system on Row Crop type tractors.

damaged hoses and/or connections.
3. Insufficient pump pressure. Check the pressure as in paragraph 71.
4. Faulty work cylinder.

BINDING. A binding or sticking condition when steering wheel is turned could be caused by:
1. Damaged or excessively worn sleeve and flange assembly at front end of control valve.
2. Binding control valve spool. Refer to paragraph 73.

EXCESSIVE FREE PLAY could be caused by:
1. Damaged or excessively worn sleeve and flange assembly at front end of control valve.
2. Improper adjustment of the control valve stop screw. Refer to paragraph 73.

HARD STEERING. This could be noticed through the steering range and should not be confused with a binding condition which usually occurs during only a portion of the steering wheel travel. Hard steering could be caused by:
1. Insufficient pump pressure. Refer to paragraph 71.
2. Faulty work cylinder and/or control valve. Refer to paragraphs 73 and 74.

NOISY PUMP OPERATION could be caused by:
1. Insufficient fluid in system.
2. Faulty or improperly adjusted pump drive belt.
3. Damaged pressure hose.

STEERING CHATTER could be caused by:
1. Bent or loose work cylinder rod.
2. Cylinder rod loose on support, or support loose on tractor side rail.

3. Faulty or improperly adjusted pump drive belt.

UNEQUAL TURNING RADIUS could be caused by:
1. Improperly adjusted linkage. Refer to paragraph 70.
2. Improper meshing of pedestal bevel gears. Refer to paragraph 20 or 22.
3. Improperly assembled main steering gear unit. Refer to paragraph 31.

LINKAGE ADJUSTMENT

70. To check and/or adjust the power steering operating linkage, refer to Fig. FO48 and proceed as follows: Measure the center to center distance between ball stud (S) and front drag link end (X). If the measured distance is not 5⅛ inches as shown, loosen clamp (C), disconnect the front drag link end from steering arm and turn drag link end either way as required to obtain the 5⅛-inch measurement. Reconnect drag link end to steering arm and tighten clamp (C).

Measure the center to center distance between ball stud (S) and rear drag link end (Y). If the measured distance is not 44⅝ inches as shown, loosen the clamp bolts in the drag link sleeve and turn the sleeve either way as required to obtain the 44⅝-inch measurement. Tighten the sleeve clamp bolts.

SYSTEM OPERATING PRESSURE

71. A pressure test of the hydraulic circuit will disclose whether the pump or some other unit in the system is malfunctioning. To make such a test, proceed as follows: Connect a pressure gage and shut off valve in series

with pump pressure line as shown in Fig. FO49. Notice that the shut-off valve is connected in the circuit between the pressure gage and the control valve.

Run the engine at slow idle speed until the oil is warmed to normal operating temperature; then turn the front wheels to the extreme right or extreme left position. With the shut-off valve open, the gage should read approximately 1100 psi if equipped with rotor type pump or 700-800 psi if equipped with hollow plunger or roller vane type power steering pump. Note: To avoid heating the fluid excessively, don't hold the wheels in this test position for more than 45 seconds.

If the pressure is higher than specified, the relief valve in pump cover is probably stuck in the closed position. If the pressure is lower than specified turn the front wheels to the straight ahead position; at which time, the gage pressure should drop considerably.

Now slowly close the shut-off valve and retain in the closed position only long enough to observe the pressure gage reading. Note: Pump may be seriously damaged if valve is left closed for more than 15 seconds. If the gage reading increases to 1100 psi on rotor type pump or to 700-800 psi when equipped with hollow plunger or roller vane pump, the pump and relief valve are O.K. and any trouble is located in the control valve or power cylinder. If the gage pressure is too low, renew the relief valve and/or spring and recheck the pressure reading. If a new relief valve and spring does not restore the system pressure, it will be necessary to overhaul the pump as in paragraph 76, 78 or 80.

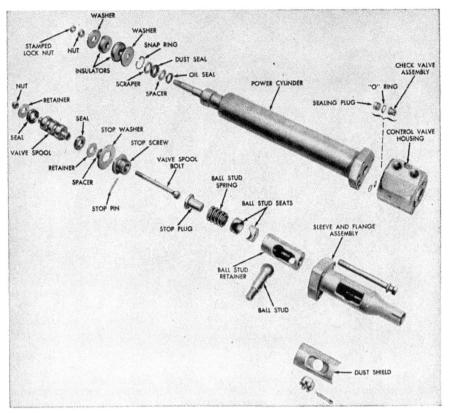

Fig. FO50 — Exploded view of Row Crop power steering cylinder, valves and associated parts.

PUMP AND RESERVOIR

Three different types of power steering pumps have been used on rowcrop tractors. Refer to paragraph 77 for early rotor type pump, to paragraph 75 for later hollow plunger type pump, or to paragraph 79 for latest roller vane type pump.

CONTROL VALVE AND POWER CYLINDER ASSEMBLY

72. **REMOVE & REINSTALL.** Disconnect hoses at control valve and secure ends of same above pump reservoir to prevent fluid drainage. Turn front wheels full right and full left several times to force all the oil from power cylinder and control valve. Remove cotter key and castellated nut from ball stud and using a suitable puller disconnect drag link from sleeve housing. (CAUTION: Do not use wedge or pinch bar to disconnect drag link as serious damage could result to sleeve housing or valve housing). It is not necesary to remove drag link from steering gear pitman arm. Unbolt and lift cylinder and valve assembly from tractor.

After cylinder and valve are reinstalled, fill and bleed the system as in paragraph 68.

73. **OVERHAUL CONTROL VALVE.** Place the power cylinder and control valve assembly in a vise, remove the retaining bolts and separate the cylinder and control valve. Refer to Fig. FO50. Remove the locknut from the end of the control valve spool bolt and remove the sleeve and flange assembly from the control valve housing. Remove retainers, seals and control valve spool from valve housing. Remove the sealing plug by inserting a $\frac{3}{32}$-inch Allen wrench into the valve housing oil return hole and forcing the plug out of cylinder end of housing. Unscrew the check valve.

Place the sleeve and flange assembly in a vise and pull the valve spool bolt outward until the ball stud is against the end of the elongated slot of the sleeve and remove the stop pin from the valve stop screw. Unscrew the valve spool stop screw from the ball stud retainer and remove valve spool bolt and valve spool stop screw. Remove the stop plug, spring and one ball stud seat; then slide the ball stud and retainer toward the closed end of the sleeve and remove ball stud by dropping same into the retainer and out the open end.

Clean all parts and inpsect for wear or damage. Remove any burrs from control valve spool and control valve housing with crocus cloth, however, DO NOT round of any sharp edges of the control valve spool as it may affect its operation. The control valve spool should fall freely of its own weight when placed in the control valve housing. (Specified clearance is 0.0002-0.0005.)

Reassembly is basically the reverse of disassembly, however, these points should be kept in mind. Reassemble the sleeve and flange assembly first and when installing the stop screw, tighten it securely; then, back it off until the nearest hole in the stop screw is aligned with the hole in the head of the valve spool bolt head. Coat the control valve parts with Ford hydraulic oil before installation. Be sure that the larger seal journal end of the control valve spool points toward the sleeve and flange end of the control valve housing and that the seal retainers are installed with their diameters matching those of the control valve spool ends. Before joining sleeve and flange assembly to control valve housing, insert top mounting bolt as there will be the necessary clearance at this time. After mating the two assemblies, install the locknut on the control valve spool bolt and tighten the nut enough to seat the parts solidly; then, loosen the nut sufficiently to allow the spool to rotate freely in the housing with a minimum of end play. The balance of the reassembly is self evident.

74. **OVERHAUL POWER CYLINDER.** The construction of the power cylinder does not allow for repair of piston or piston rod. If internal failure occurs, the cylinder assembly must be renewed as a unit. However, seals on piston rod can be renewed if they become worn or damaged.

Thoroughly clean cylinder, remove snap ring from piston rod and pull rod outward to remove dust seal and scraper. Use a sharp instrument and pry the seal from its seat. Lubricate the oil seal and install same with the cup toward the piston. Install the spacer, dust seal and scraper using an $\frac{11}{16}$-inch deep wall socket as a driver. Install snap ring.

NOTE: Power steering cylinder can be cleaned internally by removing control valve assembly, placing flanged end of cylinder in container of petroleum solvent, and working the piston rod to pump solvent into and out of the cylinder. Drain solvent from cylinder and lubricate with Ford M-2C-41 oil prior to reassembling.

POWER STEERING PUMPS

HOLLOW PLUNGER TYPE PUMP AND RESERVOIR

NOTE: This section covers the hollow sleeve plunger type pump used on general purpose and utility type tractors prior to Serial No. 89929 and row crop tractors between Ser. Nos. 5808 and 89929. Refer to paragraph 79 for roller vane type pump used on all tractors after Ser. No. 89929, and to paragraph 77 for rotor type pump used on row crop tractors prior to Ser. No. 5808. (All serial numbers apply to 1958-62 production.)

75. REMOVE AND REINSTALL. Disconnect pressure line from top of pump body. Disconnect return line from rear of reservoir and cap the pump reservoir inlet. Loosen turnbuckle and remove pump drive belt, then remove clevis pin retaining turnbuckle to pump. Remove bolt from pump mounting bracket and lift pump from tractor.

Reinstall by reversing the removal procedure, then refill and bleed system as outlined in paragraph 37 or 68.

76. OVERHAUL. Remove pump from tractor as outlined in paragraph 75 and drain oil from reservoir. Cap off discharge and return line fittings and thoroughly clean pump. Loosen clamp ring and remove reservoir from pump body. Remove cotter pin from hex head fitting and remove filter assembly from reservoir. Remove cap screw retaining drive pulley to pump shaft and using a suitable puller remove pulley from pump shaft, then remove Woodruff key.

Mount pump in a vise and loosen the four cap screws at the inlet end cap (Fig. FO51). Unscrew spring retainer and using care to prevent flow control spring from flying out, remove spring, screen and flow control valve.

Reverse pump in vise and loosen the five cap screws in bearing cap end, then remove pump from vise and remove both end caps. Remove the drive cylinder block, the driven cylinder block, the nine bent sleeves and the cylinder plunger and spring from the pump body.

Use Tru-Arc No. 3 snap ring pliers, or equivalent, and remove snap ring from bearing end cap, then push bearing and shaft assembly from bearing cap.

Clean all parts **except** the sealed ball bearing in a suitable solvent. If bearing or shaft are renewed, install

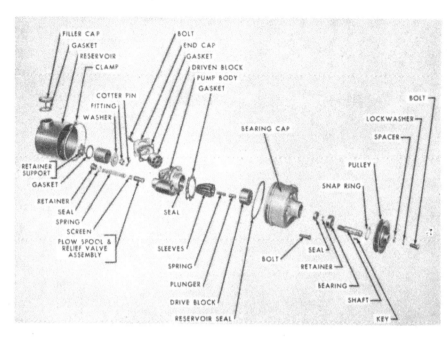

Fig. FO51 — Exploded view of hollow sleeve type pump used on early All-Purpose and Row Crop tractors. Refer to Fig. FO62 for roller vane type pump, and Fig. FO58 for rotor type pump.

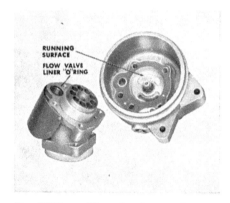

Fig. FO52 — View showing running surfaces of the bearing end cap. Surfaces should be free of gouging or scratches. Note also the "O" ring seal in flow control valve bore.

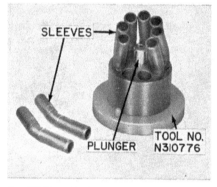

Fig. FO54 — Position driven cylinder block, plunger and sleeves as shown before installing pump body.

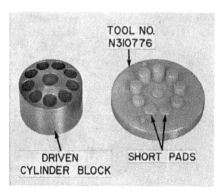

Fig. FO53 — Note position of the two short pads of assembly fixture (tool No. N310776) prior to positioning driven cylinder block on fixture.

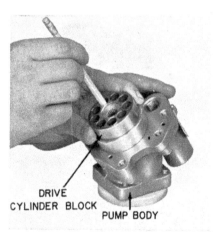

Fig. FO55 — With pump body and drive cylinder block positioned, align sleeves using method shown.

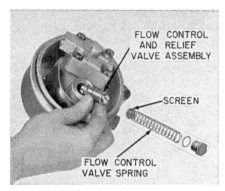

Fig. FO56 — Install flow control and relief valve in pump body with slotted end up (rearward) as shown. Note screen in end of flow control valve spring.

bearing with trade-marked side toward pulley end of shaft and press bearing on shaft until it bottoms against retaining ring. The retaining ring is located between bearing and splined end of pump shaft and must not be worn or distorted.

Check to see that the nine bent sleeves slide freely in the bores of the cylinder blocks. Inspect the mating surfaces of sleeves and cylinder blocks. Heavy scoring can impair pumping efficiency and if this condition is found, either the sleeves or cylinder blocks, or both, must be renewed. Check plunger for freedom of movement in driven cylinder block and test plunger spring which should exert a force of 29.7-36.3 lbs. when compressed to a height of 61/64-inch.

Unscrew relief valve seat from flow control valve assembly and withdraw ball, plunger, and spring. Examine all parts of both the relief valve and flow control valve. See that flow control valve moves freely in its bore. Small hair-line scratches are permissible in flow control valve; however, if burrs and scratches are found which tend to cause valve to stick, the valve should be renewed. Flow control valve spring should test 11.25-13.75 lbs. @ 1½-inches. If relief valve and/or spring show signs of damage it is recommended by the Ford Motor Company that the entire flow control and relief valve assembly be renewed as a unit.

When installing a new shaft seal, the lips of the seal must point inward and seal should be pressed in until it bottoms. Examine the running surfaces of the bearing end cap and the inlet end cap (See Fig. FO52) for heavy scratches or gouging and if these are in evidence, or if part edges are damaged, renew the part.

Filter can normally be reused after cleaning; however, if filter is cracked or shows signs of being plugged, renew the element.

When reassembling, use new "O" rings and gaskets and proceed as follows: Install bearing and shaft assembly by pressing on outer race of bearing. Use care not to mar bronze bushing. NOTE: Bronze bushing is not serviced separately and if defective renew bearing end cap. After bearing is positioned, install snap ring and check rotation of shaft.

Place driven cylinder block (without splines) on assembly fixture (Nu-day Tool No. N310776) and note the position of the two shortest pads in fixture. See Fig. FO53. Install spring and plunger in center hole. Coat the bent hollow sleeves with light oil and insert seven of them to fill all holes EXCEPT those two over the short pads as shown in Fig. FO54. Place body of pump over cylinder block, square end down, and position pump body on locating pin of assembly fixture. Insert the remaining two bent hollow sleeves and space all sleeves uniformly. Position drive cylinder block (with splines) over the bent hollow sleeves and working through holes in the drive cylinder block, align the bent hollow sleeves until the drive cylinder block will move into place. CAUTION: Do not use force during this operation as the drive cylinder will move smoothly into place when alignment is correct. Refer to Fig. FO55.

Install a new "O" ring in the counter-bore of the flow control valve bore, then, using a new gasket install bearing end cap so that locating pin in bearing end cap engages hole in pump body. Install cap screws finger tight only. Lift pump body and cylinder block unit from assembly fixture, being careful not to allow plunger to push cylinder blocks from pump body. Using a new gasket, install inlet end cap and tighten cap screws finger tight only. Now check pump to see that it rotates freely, with no binding. If pump rotates freely, tighten end cap screws uniformly and to a torque of 15-20 Ft.-Lbs. Note: There is little possibility that pump rotation will not be satisfactory; however, if such is the case, the complete assembly procedure will have to be rechecked.

Mount pump assembly in a vise in a vertical position. Install relief valve assembly in flow control valve and install the complete assembly into its bore with the slotted end up as shown

in Fig. FO56. Insert flow control valve spring with screen end down and using a new "O" ring, install and tighten retainer until "O" ring is seated.

The balance of assembly is evident, but keep in mind that the pulley should not be drawn so far on the shaft that it bears against the bearing. To do so would result in bearing failure. Also, do not attempt to press pulley onto shaft as this will only press shaft into pump, unseat front drive block from bearing cap and cause pump malfunction and/or damage. Adjust belt tension to allow ½-inch deflection.

ROTOR TYPE PUMP AND RESERVOIR

NOTE: This section covers rotor type pump used on row crop tractors prior to Ser. No. 5808. Refer to paragraph 75 for hollow sleeve type pump used between Ser. Nos. 5808 and 89929 and paragraph 79 for roller vane type pump used on all models after Ser. No. 89929. (All serial numbers apply to 1958-62 production.)

77. REMOVE AND REINSTALL. To remove the power steering pump proceed as follows: Remove the reservoir cover and withdraw as much oil as possible from the reservoir with a suction gun. Disconnect hoses from pump pressure and return line fittings and secure the hoses in a raised position to prevent oil drainage. Loosen the pump drive belt adjusting bolts and remove belt. Withdraw bolts and lift pump, reservoir and adjusting bracket off as a unit.

After pump is installed, fill and bleed the system as in paragraph 68.

78. OVERHAUL. Before disassembly make provision to prevent entry of dirt and foreign material into pressure and return fittings, and thoroughly clean pump and reservoir assembly. Clamp adjusting bracket in a vise, unhook filter retaining spring and lift filter element from reservoir. Remove reservoir retaining bolts and remove filter support plate, reinforcement plate and reservoir from pump body. Remove the drive pulley and key from pump shaft; then unbolt and separate the pump cover from the pump body (Tap gently if necessary). Remove the pump rotors and the rotor drive key from the rotor shaft and mark rotors with chalk so they can be reassembled in the same

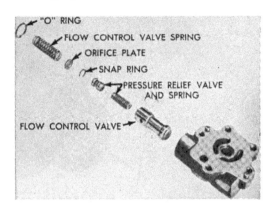

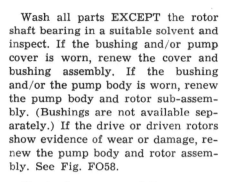

Fig. FO57 — Components of combination flow control and relief valve used on rotor type pump.

Wash all parts EXCEPT the rotor shaft bearing in a suitable solvent and inspect. If the bushing and/or pump cover is worn, renew the cover and bushing assembly. If the bushing and/or the pump body is worn, renew the pump body and rotor sub-assembly. (Bushings are not available separately.) If the drive or driven rotors show evidence of wear or damage, renew the pump body and rotor assembly. See Fig. FO58.

Check clearances as follows: Insert rotor shaft and bearing into housing until bearing is in position. Install key, drive rotor and driven rotor and check clearance at tooth ends as shown in Fig. FO59. If clearance exceeds 0.006 renew pump body and rotor assembly. Check clearance between top of rotors and surface of pump body with a straight edge and feeler gage as shown in Fig. FO60. If clearance exceeds 0.0025 renew pump body and rotor assembly. Check clearance between driven rotor and insert in pump body as shown in Fig. FO61. If clearance exceeds 0.006 renew pump body and rotor assembly. The flow control valve spring should exert 14.5-17.5 pounds, when com-

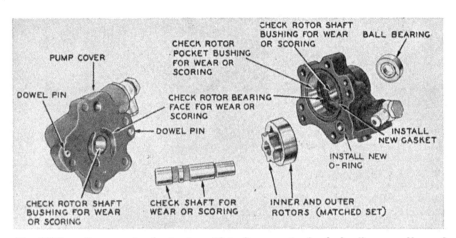

Fig. FO58 — Components of rotor type (Eaton) power steering hydraulic pump. None of the bushings are available separately.

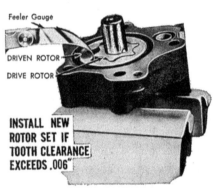

INSTALL NEW ROTOR SET IF TOOTH CLEARANCE EXCEEDS .006"

Fig. FO59 — If a 0.007 feeler blade can be inserted as shown, install a new pump body and rotors assembly.

Fig. FO60 — If a 0.0035 feeler blade can be inserted between faces of rotors and underside of straight edge, install a new pump and rotor assembly.

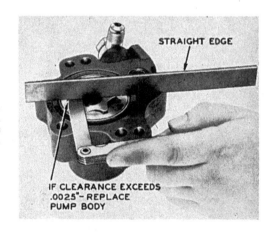

position. Remove the rotor shaft bearing retainer and press the rotor shaft and bearing from the pump body. Remove the pump outlet connection from the cover and withdraw the flow control spring, orifice plate and flow control valve. Remove the snap ring retaining the pressure relief valve in the flow control valve and remove the pressure relief valve and spring. Refer to Fig. FO57.

Fig. FO61 — If a 0.007 feeler blade can be inserted between driven rotor and pump body insert, install a new pump body and rotors assembly.

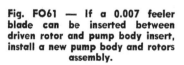

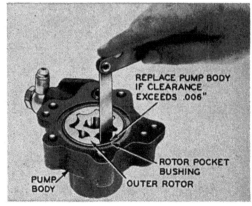

pressed to a height of 1 13/64 inches. The pressure relief valve spring should exert 21-23 pounds when compressed to a height of 1 3/16 inches. Renew springs if they do not meet specifications.

Thoroughly dry relief valve and bore of flow control valve; then insert relief valve and make sure it moves freely in bore of flow control valve. If necessary to remove any burrs, use crocus cloth. Check freedom of movement of flow control valve in pump cover bore in the same manner.

Before assembly coat all parts with clean, Ford Hydraulic Fluid M2C41. If a new rotor shaft seal is used, coat lip of same with Lubriplate or its equivalent, and be sure it is installed with lip toward pump rotors. Always use new "O" rings. Reassemble by reversing the disassembly procedure.

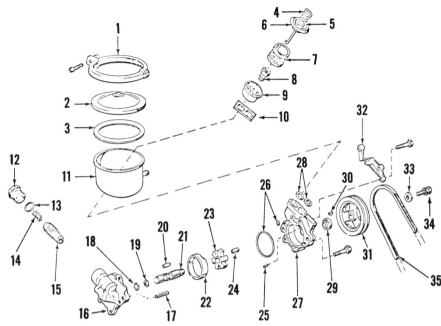

Fig. FO62 — Exploded view of roller vane type power steering pump. Pump shown is for Heavy Duty Industrial tractor; however, pump used on other type tractors is similar.

1. Cover clamp	10. Reinforcement	18. Snap ring	27. Pump body
2. Reservoir cover	11. Reservoir	19. Thrust bearing	28. Square cut rings
3. Cover gasket	12. Cap	20. Drive pin	29. Shaft seal
4. Retainer hook	13. "O" ring	21. Pump shaft	30. Woodruff key
5. Coil spring	14. Flow control spring	22. Cam ring	31. Drive pulley
6. Washer	15. Flow control &	23. Roller carrier	32. Pump bracket
7. Filter element	relief valve assy.	24. Rollers	33. Flat washer
8. Hollow stud	16. Pump cover	25. Cam retaining pin	34. Pulley cap screw
9. Filter base	17. Dowel pin	26. "O" rings	35. Drive belt

ROLLER VANE TYPE PUMP AND RESERVOIR

NOTE: This section covers the roller vane type pump used on all tractors after Ser. No. 89929 of 1958-62 production. For earlier rotor type pump refer to paragraph 77, and for hollow sleeve plunger type pump refer to paragraph 75.

79. REMOVE AND REINSTALL. To remove the pump and reservoir as an assembly, first withdraw as much oil as possible from the reservoir with a suction gun. Loosen turnbuckle and disconnect drive belt. Remove clevis pin from turnbuckle and pump. Loosen cap screw retaining pump bracket to cylinder head. Remove cap screw which secures pump bracket to water outlet connection. Disconnect pressure and return hoses from pump assembly. Fasten hoses in a raised position to prevent oil drainage. Remove pump, reservoir and bracket from tractor. Reinstall by reversing the removal procedure.

80. OVERHAUL. To overhaul the pump, refer to Fig. FO62 and proceed as follows:

Clamp the pump support in a vise and remove filter screen and component parts from reservoir. Remove the reservoir and "O" ring seals from top of pump. Remove pulley from pump shaft. Remove cap screws holding cover and bracket to pump body and lift cover vertically to prevent internal parts from falling out. Remove the two "O" ring seals from pump

body and withdraw the pump shaft. Always remove shaft toward cover end of pump body. Lift out carrier, rollers, cam ring and retaining pin from pump body. Remove shaft seal from pump body, taking care not to damage shaft bushing. The system pressure relief valve is contained within the flow control valve and component parts are not serviced separately. Therefore, if relief valve is faulty, renew the complete flow control and relief valve assembly.

NOTE: When renewing the pressure relief and flow control valve assembly, be sure to use the correct Ford part number according to type of tractor pump is used on; otherwise, malfunction and/or damage of power steering system may result. Correct part numbers are as follows:

Part No. B8NN-3A561-A (700-800 psi) should be used on All-Purpose, Rowcrop, Utility Industrial, LCG and Grove type tractor power steering pumps.

Part No. CONN-3A561-C (1350 psi, max.) should be used only on Heavy Duty Industrial models with drag link (power assist) steering.

Part No. 313366 (1550 psi, max.) should be used only on Heavy Duty Industrial models with full power steering.

Wash all parts in a suitable solvent and dry with compressed air. Check pump body and cover and renew if inner surfaces are scored or worn. Shaft bushings are not serviced separately, if worn, the body and/or cover will have to be renewed. Use Nuday Tool No. 33623 or a socket of a size that will apply pressure to outside edge of metal portion of seal and install double lip oil seal (Fig. FO63) with metal portion of seal facing outside of pump as shown. Inspect cam retainer pin and renew if bent or worn. Renew cam ring if worn or damaged. Install cam with slot over the retaining pin. Install protector sleeve (Nuday Tool No. 33623-B) or shim stock in seal and insert shaft from rotor side of pump body forcing out sleeve or shim stock as shaft enters seal. Install carrier on shaft, and position the retaining snap ring. See Fig. FO64 for proper installation of carrier. Inspect rollers and renew if scored, damaged or out-of-round.

Fig. FO63 — **Proper method of installing shaft oil seal in roller vane power steering pump.**

Fig. FO64 — **Install roller carrier in pump so that it will turn in the direction indicated by the arrow. See text.**

Check end clearance of the carrier and rollers by placing a straight edge across the body and measuring clearance with a feeler gage. If end clearance exceeds 0.0025, renew the carrier and rollers. Inspect flow control valve assembly and valve bore; remove all burrs with crocus cloth.

Insert flow control valve in its bore so that extension enters first, as

shown in Fig. FO65. Install new "O" ring and tighten valve cap to a torque of 30-35 Ft.-Lbs. Install new "O" rings, be sure thrust spacer is in place in bearing bore of pump cover, place pump body and cover together and secure loosely with the three short cap screws; then, install pump bracket on the pump. Tighten all five cap screws evenly to a torque of 20-25 Ft.-Lbs. Check pump shaft for freedom of rotation, then install reservoir and tighten retainer bolt to a torque of 30-35 Ft.-Lbs. Filter element is held seated in reservoir by a spring loaded hook. The spring and seat acts as a relief valve, allowing oil to bypass the filter if filter becomes clogged. Reinstall filter, fasten pulley on pump shaft with retaining cap screw and tighten to a torque of 15-20 Ft.-Lbs. Pulley is installed with flat side toward pump body. Reinstall pump assembly on engine and adjust drive belt to ¼ to ½-inch deflection by means of the turnbuckle. Fill the reservoir with Ford hydraulic oil, M2C41 to within ½-inch of top and cycle system several times to bleed air from the cylinder and lines.

NOTE: Refer to paragraph 60 for special bleeding procedure required on Heavy Duty Industrial models with drag link (power assist) steering. Recheck the fluid level and refill as required.

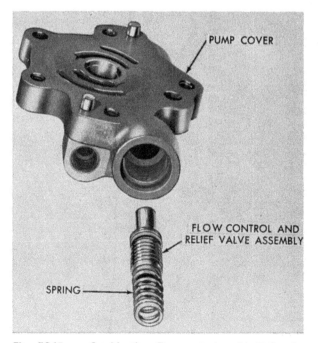

Fig. FO65 — **Combination flow control and relief valve assembly is inserted into pump body small end first as shown.**

ENGINE AND COMPONENTS

Gasoline and LP-Gas versions of Series 501, 600, 601, 700, and 2000 tractors are equipped with an engine having a bore of 3.44 inches, a stroke of 3.60 inches and a piston displacement of 134 cubic inches. Diesel versions of the above tractors are equipped with an engine having a bore of 3.56 inches, a stroke of 3.60 inches and a piston displacement of 144 cubic inches. Series 800, 801, 900, 901, 1801 and 4000 tractors are equipped with a gasoline, LP-Gas or Diesel engine having a bore of 3.90 inches, a stroke of 3.60 inches and a piston displacement of 172 cubic inches. The basic design of all engines (including diesels) is similar and many of the component parts are interchangeable.

R&R HOOD ASSEMBLY AND FUEL TANK

To remove the cylinder head from engine or to remove engine from tractor, the hood assembly and fuel tank must first be removed from all models as outlined in the following paragraph.

81. First, disconnect ground cable from battery; then, proceed as follows:

On "Select-O-Speed" transmission models with power take-off, carefully drive the roll pin out of PTO control knob, remove the knob and then remove the hex nut and washer that retain the PTO cable housing in the hood panel.

On non-diesel models, remove the air cleaner inlet screen, or pre-cleaner from left hood side panel, then remove the air cleaner "funnel" from opening in hood panel.

On early diesel models with air cleaner attached to intake manifold, remove the pre-cleaner air stack, if so equipped, from tractor.

On diesel models with air cleaner located at left side of hood, disconnect the rubber hose at intake manifold inlet pipe, remove the two top starter mounting bolts, remove the bolts attaching air cleaner bracket to hood and lift the air cleaner, pre-cleaner and air tube from tractor as a unit.

On 600, 601 and 800 Series and early (prior to 1963) 2000 Utility Industrial models, unscrew the two wing screws at top of grille, tip grille forward and remove the two nuts,

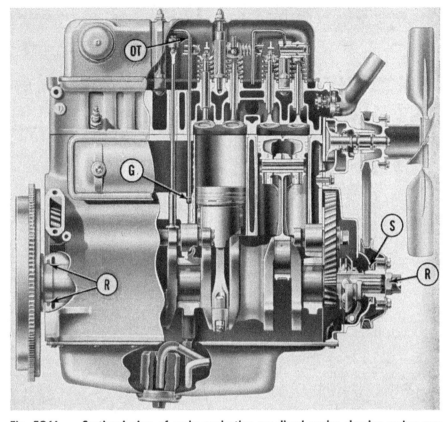

Fig. FO66 — Sectional view of early production non-diesel engine showing rocker arm oil tube (OT) and grommet (G), crankshaft pulley retaining nut (N), front crankshaft seal (S) and rear crankshaft seal (R). Called out parts of diesel engine are similar.

flat washers and insulator washers that retain grille to hood lower side panels. Unbolt hood lower side panels from front support and remove grille.

On 801 Series and early (prior to 1963) 4000 Utility Industrial models, remove the grille, remove lower front panel from hood lower side panels and unbolt hood lower side panels from front support.

On 1963 and later Series 2000 and 4000 All-Purpose, Utility Industrial and LCG models, remove the grille, tip lower grille panel forward and remove the two nuts, flat washers and insulator washers that retain lower panel to hood lower side panels. Unbolt hood lower side panels from front support and remove lower grille panel.

On all Row Crop and Offset models, remove the grille and unbolt hood lower side panels from pedestal or front support. On Offset models only, disconnect steering shaft universal joint from steering gear shaft, unbolt

steering shaft bracket from top of transmission and remove steering wheel, bracket and shaft as a unit.

On Heavy Duty Industrial models, disconnect headlight wire at connector in front of radiator and unbolt front end of hood from cast iron radiator grille.

On all models, disconnect headlight wire at connector near the headlight switch on right side of rear hood panel. Remove the screws, nuts and cap screws that retain the hood side panels to instrument panel and rear hood panel. Be sure the "Select-O-Speed" PTO control cable housing is free of hood panel. Then, lift hood assembly from tractor.

On power steering equipped models, remove the power steering pump from the front fuel tank bracket without disconnecting the power steering fluid lines.

On diesel models, remove battery from tractor if engine is to be removed; otherwise, place a board between the fuel tank and battery to insulate tank from battery terminals. Then, drain the fuel tank, remove fuel supply line to fuel filter and remove the excess fuel (fuel return) lines from injection pump and elbow fitting on cylinder head to standpipe at right front corner of fuel tank.

NOTE: On early diesels, excess fuel line from injection pump ran to a "T" fitting on cylinder head, and a single return line led from the "T" fitting to the standpipe in fuel tank. If one of these units is encountered, it is recommended that a Standpipe & Leak-Off Lines Kit (Ford part No. JPN-9003-A for 172 cu. in. displacement engines or JPN-9003-B for 144 cu. in. displacement engines) be installed when reassembling tractor.

On gasoline models, shut off the fuel supply valve and remove fuel supply line.

On LP-Gas models, shut off both the liquid and vapor withdrawal valves, be sure all pressure is bled from lines, and disconnect the hose from vaporizer. See "CAUTION" preceding paragraph 119.

Unbolt the fuel tank brackets from cylinder head, disconnect fuel gage wire from sender switch on models with fuel gage and lift the fuel tank, heat baffle and brackets from the engine as an assembly.

R&R ENGINE AND CLUTCH ASSEMBLY

All Purpose, Utility Industrial and LCG Types

82. To remove engine and clutch assembly, first drain hydraulic sump on models equipped with hydraulic system and/or power take-off; drain cooling system and, if engine is to be disassembled, drain the oil pan. After the hood and fuel tank are removed as outlined in paragraph 81, proceed as follows:

On power steering models, disconnect the spark plug wires on non-diesel engines or remove the fuel injection pump from diesel engine as outlined in paragraph 140. Then, disconnect pump tubes from control valve at steering gear housing and remove the power steering pump and the tubes as a unit.

Disconnect Proof-Meter cable (if so equipped) from hydraulic pump or drive unit. If equipped with hydraulic system, remove hydraulic pump and the pump manifold as a unit. Remove water temperature gage bulb from cylinder head and disconnect oil pressure gage tube from fitting on cylinder block.

On non-diesel models, remove choke rod and disconnect governor rod from bellcrank on transmission. Remove air tube from between air cleaner and carburetor. Disconnect wiring from generator and distributor, detach from retaining clips on engine and move wiring back out of way. Remove exhaust pipe from tractor.

On diesel models, disconnect throttle linkage from bellcrank by working through hole in bottom of battery tray. Loosen the two rear battery tray bolts, tip tray back up out of way and tighten the bolts. Disconnect wiring from generator, detach from retaining clips and move wiring back out of way. Detach wire to glow plugs on intake manifold. Remove exhaust pipe and bracket as a unit and, on early models, remove air cleaner from intake manifold.

On all models, remove oil filter, disconnect cable from starter motor and remove starter motor from tractor. Disconnect radiator hoses, place support under front end of transmission and disconnect drag link from steering arm on right front wheel spindle. Remove pin retaining right radius arm to front axle. Note: The cylinder tubes do not need to be disconnected on power steering models. Remove the nuts retaining front support to engine and pivot the front axle forward around the left radius rod to front axle pin. Attach a hoist to engine so that engine and clutch assembly will be balanced and unbolt engine from transmission. Pull the engine forward and move to work area.

Row Crop and Offset Types

83. To remove engine and clutch assembly, first drain the hydraulic system, cooling system and, if engine is to be disassembled, drain the oil pan. Remove the hood and fuel tank as outlined in paragraph 81; then, proceed as follows:

On Row Crop models, remove the power steering pump, hoses, cylinder and steering drag link as a unit.

Support tractor under front end of transmission and remove the right hand side rail from tractor. Disconnect radiator hoses and remove radiator. Be sure that pedestal or front support is adequately supported, unbolt pedestal or front support from engine and left side rail from transmission. Move the front unit away from tractor.

Disconnect Proof-Meter cable at

hydraulic pump, then remove the pump and pump manifold as a unit. Remove oil filter, disconnect starter cable and remove starter motor. Remove water temperature gage bulb from cylinder head and disconnect oil pressure gage line from fitting on cylinder block.

On non-diesel models, remove choke rod and disconnect governor rod from bellcrank on transmission. Remove air tube from between air cleaner and carburetor. Disconnect wiring from generator and distributor, detach from retaining clips and move wiring back out of way. Remove exhaust pipe from tractor.

On diesel models, disconnect throttle linkage from bellcrank by working through hole in bottom of battery tray. Loosen the rear battery tray bolts, pivot tray up as far as possible and tighten the bolts. Disconnect wiring from generator and retaining clips on engine and move wiring back out of way. Disconnect wire to glow plugs on intake manifold. Remove the exhaust pipe and bracket as a unit and, on early models, remove air cleaner from intake manifold.

Attach hoist to engine so that engine and clutch assembly will be balanced, unbolt engine from transmission flange and move engine forward from tractor.

Heavy Duty Industrial Models

84. Remove backhoe from tractor if so equipped. The Ford H.D. Industrial loader does not need to be removed. Following procedure assumes that loader is on tractor; disregard reference to loader components if not so equipped. Also, some Heavy Duty Industrial models may not be equipped with a hydraulic package (front mounted pump and reservoir).

To remove the engine and clutch assembly, proceed as follows: Drain cooling system, hydraulic reservoir on models with 3-point hitch and/or PTO and, if engine is to be disassembled, drain oil pan. It is not necessary to drain the power steering system or the hydraulic reservoir mounted in front of radiator. Remove the hood and fuel tank as outlined in paragraph 81.

Remove the two diagonal braces from loader frame to steering housing on tractor. The right diagonal braces does not need to be removed from the control valve handles. Unbolt and remove the control valve support brace (A—Fig. FO67) from the control valve, hydraulic oil filter bracket and engine. Note: On models with re-

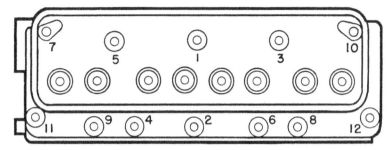

Fig. FO68 — Cylinder head hold down screws of non-diesel engines should be tightened in this numbered order and to a torque of 65-70 Ft.-Lbs.

Fig. FO67 — Power package valve bracket (A) can be disconnected from clutch housing flange and moved to the side without disconnecting the hydraulic lines. Remove the brace (A) when removing engine from tractor equipped with loader.

versing transmission, unbolt reversing lever bracket to allow clearance for unbolting the left diagonal loader brace from steering housing.

From rear end of tractor, place a rolling floor jack under front end of transmission and place supports under each side frame member near the rear wheels. Disconnect the power steering actuator cable from steering housing on models with full power steering, or detach drag link from steering gear arm on models with power assist steering. Unbolt and remove the two transmission to side frame braces and the step plates from tractor. Unbolt and remove both rear fenders after disconnecting tail light wire.

Disconnect the radiator hoses and remove the two bolts from engine and front support. On "Select-O-Speed" transmission models, disconnect the traction coupling as outlined in paragraph 223. Roll the engine, transmission and final drive assembly to rear. It may be necessary to manipulate such items as prying the hydraulic tubes outward, moving the right diagonal brace around the control valve handles, raising or lowering the engine, etc., to gain clearance as the rear unit is rolled back. Most mechanics prefer to move the engine back clear of the tractor and loader frame before detaching engine from transmission; however, engine can be removed after rear unit is rolled back about 11 inches if floor space is limited.

Disconnect Proof-Meter cable at hydraulic pump or drive unit and re-

move the hydraulic pump and pump manifold on models equipped with a 3-point hydraulic lift system. Remove engine oil filter, disconnect starter motor and remove starter motor from tractor. Disconnect wiring harness from generator and, on non-diesel models, the distributor; then detach wiring from retaining clips on engine and move wiring back out of way. On diesel models, disconnect wire to glow plugs on intake manifold and, on early models, remove air cleaner from intake manifold.

On non-diesel models, remove choke rod and disconnect governor rod from bellcrank on transmission. Remove air tube from between air cleaner and carburetor and remove the exhaust pipe.

On diesel models, disconnect throttle linkage from the bellcrank by working through hole in bottom of battery tray. Loosen rear battery tray bolts, pivot tray upward as far as possible and reighten the bolts.

Attach a hoist to engine so that engine and clutch assembly will be balanced, unbolt engine from transmission and move engine forward.

Before reinstalling engine, remove hydraulic pump drive coupling from crankshaft pulley and slide it on the splines of pump drive shaft. Before reattaching drive coupling to crankshaft pulley, be sure that fan belt and power steering pump drive belt are in place. Refer to Fig. FO297 for installation of hydraulic pump drive shaft and adjustment of the end clearance.

CYLINDER HEAD

Non-Diesel Models

85. To remove the cylinder head, first drain the cooling system and remove hood and fuel tank as outlined in paragraph 81. Remove muf-

fler, then disconnect the air cleaner hose, choke rod and governor link from the carburetor. Remove manifold and carburetor as a unit. Remove the cylinder head water outlet casting, or the upper radiator hose, and remove temperature gage bulb from cylinder head. Remove the rocker arm cover. Loosen rocker arm shaft support cap screws alternately and evenly, and remove the assembly. Remove the oil inlet and outlet lines. If the grommet (G—Fig. FO66) is missing from lower end of oil tube, be sure to extract it using a wire hook after cylinder head has been removed. Remove the rocker arm shaft assembly. Remove the exhaust valve free spin caps from exhaust valve stems and place them in numbered trays so that they may be reinstalled in same position. Remove the push rods and place them in numbered rack. Remove the cylinder head retaining cap screws and lift cylinder head from engine.

When reinstalling the head, tighten the retaining cap screws in the sequence shown in Fig. FO68 and to a torque of 65-70 Ft.-Lbs. Tighten the rocker arm support cap screws to a torque of 45-55 Ft.-Lbs. Check the valve tappet gap, cold, as outlined in paragraph 94; then, after engine is started and at normal operating temperature, readjust the intake and exhaust valve tapet gap to 0.015 with engine running at idle speed.

NOTE: On late production non-diesel engines, a bracket (Ford part No. C3NN-6574-A) is installed under the rear center rocker arm support cap screw head (second rocker arm support from rear of engine) to retain and seat the rocker arm oil inlet line in counterbore in cylinder block, and prevent leakage at the grommet (G—Fig. FO66). It is recommended that this part be obtained and installed during reassembly of earlier production engines.

TOOL NO. DDN 17098

Fig. FO69 — Special tool DDN17098 can be used to pull injectors. If injectors are being pulled while fuel tank is on tractor it may be necessary to bend handle of tool to prevent it striking tank.

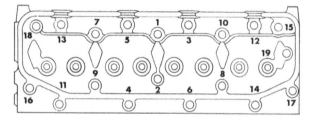

Fig. FO70 — On diesel models, the cylinder head hold down screws should be tightened in this numbered order and to a torque of 100-105 Ft.-Lbs. Number 19 also retains rear rocker arm shaft support.

Diesel Models

86. Remove the diesel cylinder head, first drain the cooling system and remove the hood and fuel tank as outlined in paragraph 81. Then, proceed as follows:

On early models, remove the air cleaner assembly from the bottom side of intake manifold.

Disconnect the wire to glow plugs on intake manifold. Remove the muffler and exhaust pipe; then, unbolt and remove the intake and exhaust manifolds. Remove the thermostat housing (water outlet) from front end of cylinder head and water temperature gage bulb at rear end of head.

Remove the valve rocker arm cover and gasket. Equally and alternately loosen each of the five rocker arm support retaining cap screws until free of all valve spring tension. Then, remove the cap screws, oil inlet line, oil return line and rocker arm assembly from cylinder head. Note: The rear rocker arm support retaining cap screw is also a cylinder head retaining cap screw. Be sure the rubber grommet is on the lower end of the oil inlet line. If not, remove the grommet from cylinder block with a

wire hook after removing the cylinder head. Remove the free spin exhaust valve caps and place in numbered tray so that they may be reinstalled in same position. Remove the rocker arm push rods and place in a numbered rack.

Remove the excess fuel (bleed back) line, disconnect injector pressure lines and remove the fuel injectors and sealing washers as outlined in paragraph 137.

Remove the cylinder head retaining cap screws and free the cylinder head by prying in the special slots provided between head and cylinder block on left side of engine. The diesel head gasket adheres firmly to both head and block, and the gasket must be split to free the head. Lacquer thinner will soften the gasket material so that it can be scraped loose easily with a sharp putty knife. CAUTION: The lacquer thinner is highly inflammable.

87. To reinstall the diesel cylinder head, proceed as follows: Install head using new gasket and tighten the retaining cap screws to a torque of 100-105 Ft.-Lbs. in order shown in Fig. FO70. Install the injectors, using new

sealing washers, and the excess fuel line as outlined in paragraph 137. Insert the push rods in their respective bores and place the exhaust valve free spin caps on their respective valve stems. Set the rocker arm assembly on the cylinder head, install new oil seal grommet on oil inlet tube and loosely install the retaining cap screws with the oil inlet and return tube brackets in place. Be sure the rocker arm adjusting screws are all seated in the push rod cups, then equally and alternately tighten the rocker arm support retaining cap screws until rocker arm supports are drawn against the cylinder head. Tighten the rear cap screw to a torque of 100-105 Ft.-Lbs. and the four front support retaining cap screws to a torque of 45-55 Ft.-Lbs.

After installing the rocker arm assembly, adjust tappet gap cold as outlined in paragraph 94. Install the intake and exhaust manifolds using a new gasket and tighten the retaining nuts and cap screws to a torque of 40-50 Ft.-Lbs. Complete the reassembly of tractor by reversing disassembly procedure, bleed the fuel injection system as outlined in paragraph 130 and start the engine. After the engine is at normal operating temperature, remove the rocker arm cover and readjust the valve tappet gap to 0.015 hot for both intake and exhaust valves: Note: To facilitate removal of rocker arm cover with fuel tank installed, remove the cover retaining studs from rocker arm supports. Tilt top of rocker arm cover out and push cover free of fuel tank with long screwdriver or punch inserted between muffler and manifold.

VALVES AND SEATS

All Models

89. Exhaust valves are equipped with free type rotators shown in Fig. FO71 and, except on 144 cu. in. diesel engines, the valves seat on renewable type seat inserts which are a shrink fit in the cylinder head. Intake valves and, on 144 cu. in diesel engines, the exhaust valves seat directly in the cylinder head. Intake valves are equipped with stem oil deflectors. Valves have a face angle of 43½ degrees and a seat angle of 45 degrees. Desired valve seat width is 0.060-0.080 for the intake, 0.090-0.110 for the exhaust. Seats can be narrowed, using 30 and 60 degree stones. Total seat runout should not exceed 0.0015 for non-diesel models; 0.002 for diesel models.

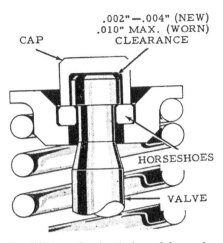

Fig. FO71 — Sectional view of free valve (release type) exhaust valve rotator. Note required end clearance between cap and end of valve stem.

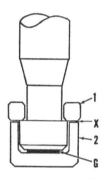

Fig. FO72 — Feeler stock method of checking end clearance between rotator cap and end of valve stem. Refer to text.

Recommended valve stem to guide diametral clearance for all valves is 0.001-0.002. A special Perfect Circle valve stem seal is available separately as a service item.

New intake valve stem diameter is 0.3415-0.3425 for non-diesel models and 0.3416-0.3423 for diesels. Exhaust valve stem diameter is 0.3405-0.3415 for all models. Renew the valves if bent, or if stems are excessively worn or scored.

VALVE GUIDES AND SPRINGS

All Models

90. Intake and exhaust valve guides are interchangeable only on non-diesel models and can be pressed from cylinder head if renewal is required. The inside diameter of a new guide is 0.3436-0.3444 for all guides. Before removing guides, measure their height above the valve spring contacting sur-

Fig. FO73 — When reassembling rocker arm assembly on diesel engines be sure spacers are located on shaft as indicated.

face of the head and install new guides to the same dimension.

A special Perfect Circle valve stem seal is available as an accessory. A step must be machined on upper outside diameter of guide for seal installation.

Valve stem diameter and running clearance is listed in paragraph 89.

Intake and exhaust valve springs are interchangeable. Renew any spring that has lost its protective coating, has rusty spots or does not test 54-62 lbs. when compressed to 1 53/64 inches or 124-140 lbs. when compressed to 1½ inches.

When reinstalling valve springs on diesel models, be sure that the ¾-inch spacer washer is installed between the exhaust valve spring and cylinder head.

Install all valve springs with the dampener coils (closed end) toward cylinder head.

EXHAUST VALVE ROTATORS

All Models

91. Refer to Fig. FO71. The free-valve type rotators will not function unless there is a measurable clearance or end gap between the end of the exhaust valve stem and the inside floor of the cap when the open end of the cap just contacts the spring keeper or horseshoe as shown. Desired end gap is 0.002-0.004. Rotator cap gaps should be checked and if necessary adjusted each time the valves are reseated.

93. One of the simpler methods of checking is shown in Fig. FO72. From a strip of 0.010 flat shim stock, cut a $\frac{3}{16}$-inch diameter disc. Lay this disc (G) which must be flat, on inside floor of rotator cap (2) and install valve lock or keeper (1) on valve stem. Now, while simultaneously pressing downward on valve lock and upward on rotator cap, measure with a feeler gauge the gap (X) between cap and valve lock. If gap measures anywhere between 0.006 and 0.009, it

is within desired limits. If gap (X) is less than 0.006, grind or lap open end face of cap; if more than 0.009, grind end of valve stem.

VALVE TAPPETS

All Models

94. Intake and exhaust valve tappet gap should be set to 0.015 when engine is at operating temperature, and with engine running at idle speed.

NOTE: When engine has been disassembled, the tappet gap should be adjusted cold before starting engine. Adjust both the intake and exhaust valve tappet gaps to 0.015 by following procedure:

With No. 1 cylinder exhaust and No. 3 cylinder intake valves open, adjust No. 2 cylinder intake and No. 4 cylinder exhaust valves.

With No. 1 cylinder intake and No. 2 cylinder exhaust valves open, adjust No. 3 cylinder exhaust and No. 4 cylinder intake valves.

With No. 2 cylinder intake and No. 4 cylinder exhaust valves open, adjust No. 1 cylinder exhaust and No. 3 cylinder intake valves.

With No. 3 cylinder exhaust and No. 4 cylinder intake valves open, adjust No. 1 cylinder intake and No. 2 cylinder exhaust valves.

The 0.4989-0.4995 diameter mushroom type tappets are available in standard size only and operate directly in the cylinder block bores with a clearance of 0.0005-0.0021. To remove the tappets, it is first necessary to remove the camshaft as outlined in paragraph 99.

ROCKER ARMS

All Models

95. To remove the rocker arms, it is recommended that hood and fuel tank be removed as outlined in paragraph 81. Remove the rocker arm cover and extract the oil inlet and outlet lines. If grommet (G—Fig. FO66) is missing

from the rear oil tube, remove the engine side plate, extract the grommet and install same over lower end of tube. Remove the cap screws retaining the rocker shaft supports to head and remove the rocker arms and shaft assembly.

All of the rocker arms are identical and interchangeable in the same type engine. The 0.780-0.781 diameter rocker arm shaft should have a clearance of 0.002-0.004 in the rocker arms.

Before assembling the rocker arm assembly, refer to Fig. FO73 to see the location of the spacers which are used on the diesel models.

When reassembling rocker arm shaft to cylinder head, tighten the rocker arm suport cap screws to a torque of 45-55 Ft.-Lbs. except the cap screw in the rear rocker arm support on diesel models. This cap screw is also a cylinder head bolt and is torqued to 100-105 Ft.-Lbs. When installing the rocker arm oil lines, be sure that grommet on lower end of the rear oil line is seated in the counterbore in the cylinder block directly above the camshaft center bearing. Adjust the intake and exhaust tappet gap to 0.015 as outlined in paragraph 94.

NOTE: On late production non-diesel engines, a bracket (Ford part No. C3NN-6574-A) is installed under the rear center rocker arm support retaining cap screw head to retain and seat rocker arm oil inlet line in counterbore in cylinder block and prevent leakage at the grommet (G—Fig. FO66). It is recommended that this part be obtained and installed during reassembly of earlier production engines.

TIMING GEAR COVER

All Models

96. To remove the timing gear cover, first drain the cooling system; then, proceed as follows:

On Offset and Row Crop models, it will be necessary to loosen the front support or pedestal attaching bolts and move the support or pedestal forward ¼ to ⅜-inch to provide clearance for removal of cover.

On Heavy Duty Industrial tractors, detach power steering pump and lay it aside without disconnecting the hydraulic lines. Then, disconnect front hydraulic pump drive hub from crankshaft pulley and remove drive hub and pump drive (universal) shaft.

On non-diesel models, disconnect governor linkage from governor arm.

On all models, remove the fan and power steering belts, unscrew crank ratchet or crankshaft pulley retaining cap screw and remove crankshaft pulley and hub from crankshaft.

The timing gear cover can then be unbolted and removed. Note: The cap screws in the front end of the oil pan are threaded into the timing gear cover. Some mechanics prefer to drain and remove the oil pan before removing the timing gear cover. If pan is not removed, the front section of a new pan gasket can be cut to fit the pan and sealed in place with Permetex or similar sealer.

NOTE: On early production non-diesel models, the governor housing was separate from the timing gear cover. On these models, the crankshaft front oil seal can be renewed and the governor serviced by removing the governor housing from timing gear cover. Neither the early type governor housing or timing gear cover are available for service; if necessary to renew either part, install new integral governor housing and timing gear cover. The gasket used between the early timing gear cover and governor housing is available. Tighten the governor housing retaining cap screw to a torque of 10-15 Ft.-Lbs.

The timing gear cover upper retaining cap screws extend into the engine water jacket. Before installing these cap screws, coat the threads with Permetex or similar sealer. Tighten the timing gear cover retaining cap screws to a torque of 10-15 Ft.-Lbs. Tighten the oil pan retaining cap screws to a torque of 20-25 Ft.-Lbs. on Offset models and 12-15 Ft.-Lbs. on other models. Tighten the crankshaft pulley retaining cap screw or ratchet nut to a torque of 100-110 Ft.-Lbs.

TIMING GEARS

All Models

97. **CAMSHAFT GEAR.** On early models where gear is retained to camshaft by a $\frac{7}{16}$ - 14 cap screw, the camshaft gear may be removed and reinstalled after timing gear cover has been removed as outlined in paragraph 96.

On late models, where gear is retained by a snap ring, camshaft must first be removed as outlined in paragraph 99.

Before removing camshaft gear or camshaft, measure the clearance between gear hub and camshaft thrust

Fig. FO74 — Camshaft gear is properly installed when punch mark on tooth space of camshaft gear is in register with punch mark on crankshaft gear tooth.

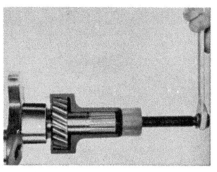

Fig. FO75 — A puller which engages rear face of gear must be used to remove crankshaft gear.

plate, using a feeler gage. Clearance should measure 0.003-0.007. If clearance is excessive, renew the thrust plate during reassembly.

Recommended timing gear backlash is 0.002-0.006. Camshaft gear is available in standard size as well as oversizes of 0.006 and 0.010 to facilitate obtaining the desired backlash. When reassembling, mesh the punch marked tooth space on camshaft gear with the punch marked tooth on crankshaft gear as shown in Fig. FO74.

98. **CRANKSHAFT GEAR.** To remove the crankshaft gear, first remove the timing gear cover as outlined in paragraph 96. On non-diesel models, withdraw governor weight assembly from front of crankshaft.

If camshaft gear must also be renewed, remove camshaft and/or gear to provide additional clearance. Attach a suitable puller as shown in Fig. FO75 to remove the crankshaft gear.

The steel crankshaft gear is keyed to crankshaft and is a press fit on shaft. When installing gear, make sure that gear face containing the timing mark is forward, and chamfer in hub of gear is towards rear.

Assembling can be facilitated by heating gear in hot oil prior to in-

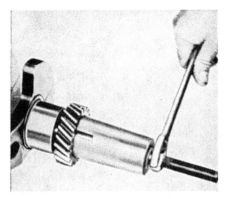

Fig. FO76 — Internally threaded front end of crankshaft permits use of stud, sleeve and nut to push gear on shaft.

Fig. FO77 — To remove pump gear from rear end of camshaft it is necessary to remove flywheel and this access plate.

stallation. Use a forcing screw and sleeve as shown in Fig. FO76, or a suitable drift. Gear bottoms on shaft shoulder when properly installed.

If camshaft gear was not removed, make sure that timing marks are aligned as shown in Fig. FO74. Use extreme care also, not to damage the teeth of the softer aluminum camshaft gear (early production non-diesel) during installation of the crankshaft gear.

Crankshaft gear is available in standard size only. Refer to paragraph 97 for recommended backlash.

CAMSHAFT

All Models

99. To remove the camshaft, first remove the engine from tractor as outlined in paragraph 82, 83 or 84, timing gear cover as outlined in paragraph 96 and rocker arms assembly and push rods as outlined in paragraph 95. Remove the clutch, flywheel and engine rear end plate. Remove the hydraulic pump drive gear access plate from rear face of cylinder block as shown in Fig. FO77 and remove the hydraulic pump drive gear from rear end of camshaft.

On non-diesels, remove the ignition distributor. On diesels, remove the injection pump as in paragraph 140 and pump drive gear as in paragraph 142. On all models, invert engine so that tappets will fall away from camshaft. Using a feeler gage, measure the camshaft end play clearance between the gear hub and the camshaft thrust plate. If the clearance is not between the limits of 0.003-0.007, renew the thrust plate during reassembly.

Working through the openings in the camshaft gear, unbolt the camshaft thrust plate from cylinder block and withdraw the camshaft from engine.

Check the camshaft and associated parts against the following values:
Camshaft journal
    diameter ....1.925 -1.926
Bearing bore diameter
    in block 1.9275-1.9285
Journal running
    clearance 0.0015-0.0035
Camshaft end play......0.003 -0.007

Reinstall the camshaft by reversing the removal procedure and tighten the hydraulic pump drive gear access plate retaining cap screws to a torque of 12-16 Ft.-Lbs., camshaft thrust plate screws to a torque of 12-16 Ft.-Lbs. and the camshaft gear retaining cap screw on early models to a torque of 35-45 Ft.-Lbs. On non-diesels, retime the ignition as outlined in paragraph 159. On diesels, install the pump drive gear as in paragraph 142, then install injection pump as in paragraph 140. Retime injection pump as in paragraph 139.

CONNECTING ROD AND PISTON UNITS

All Models

100. Connecting rod and piston units are removed from above after removing the cylinder head and oil pan. Be sure to remove top ridge from cylinder bores before attempting to withdraw the assemblies.

Connecting rod and bearing caps are numbered to correspond to their respective bores. When renewing connecting rod be sure to stamp the cylinder number on the new rod and cap.

When reassembling, make certain that the oil squirt hole in rod is toward camshaft side of engine. Dimple or the word "FRONT" in top surface of each piston must be toward front of engine. Tighten the connecting rod nuts to a torque of 45-50 Ft.-Lbs. Install new pal nuts and tighten them to a torque of 3-4 Ft.-Lbs. (or finger tight plus ⅓ turn).

NOTE: New type connecting rods having a heavier cross-section were installed as a production change in 144 and 172 cubic inch diesel engines during 1963 production. Although the new type connecting rods are slightly heavier than the rods previously used, the old and new type rods are completely interchangeable and may be installed in mixed sets without affecting engine balance.

NOTE: Effective with tractor Serial No. 82583 of 1958-62 production, a new type piston incorporating combustion chamber changes was installed in 172 cubic inch diesel engines as a production change. If necessary to renew a piston in an early production 172 cu. in. diesel engine having the original type pistons, all four pistons should be renewed.

PISTON RINGS

Non-Diesel Models

101. Each cam ground piston is fitted with two compression rings and one oil control ring. Ring sets are available in standard size as well as oversizes of 0.020, 0.030 and 0.040. New rings are marked to indicate top side and should be installed accordingly. Top compression ring is chrome plated. Both compression rings are tapered and the larger diameter of same should be installed toward bottom of piston. All rings should have an end gap of 0.010-0.020. Ring side clearance should be as follows:
Top compression ring...0.002 -0.0035
Second compression ring
    134 cu. in. engine.....0.0015-0.0035
    172 cu. in. engine.....0.002 -0.0035
Oil control ring........0.0015-0.003

Diesel Models

102. Pistons in the 144 cu. in. diesel engine are fitted with three compression rings and one oil control ring, with all rings being located above the piston pin. All four rings are chrome plated and the oil control ring is fitted with a circumferential expander spring.

The 172 cu. in. diesel engine is fitted with four rings above the piston pin of the same construction used in the 144 cu. in. engine, and also a plain cast iron oil control ring located below the piston pin at the bottom of piston skirt.

Service ring sets are similar to original production rings described in preceding paragraphs except that the top compression rings are counter-bored to accomodate wear normally found in used pistons and the third compression ring has a tapered face and is fitted with a hump type expander. Install the service ring sets according to instructions packed with each set.

Rings are available in complete sets for all four pistons only. Ring sets are available in standard size and oversizes of 0.020, 0.030 and 0.040. Use standard size ring set with 0.002 oversize pistons. All rings should have an end-gap of 0.010-0.020. Ring side clearance should be as follows:

Top compression ring

144 cu. in. engine...... 0.004 -0.006

172 cu. in. engine..... 0.0035-0.005

2nd and 3rd compression ring

144 cu. in. engine .... 0.002 -0.004

172 cu. in. engine..... 0.0015-0.0025

Oil control rings

144 cu. in. engine.... 0.0015-0.0035

172 cu. in. engine 0.0015-0.003

PISTONS AND CYLINDERS

All Models

103. Latest model engines are not equipped with sleeves and the cam ground piston operates directly in finished cylinder block bores. Older engines, except the 144 cu. in. diesel, were equipped with a semi-finished, dry, cast sleeve. The sleeves are available for service repairs on all engines.

For non-diesel engines, hand push fit cylinder sleeves are available for service installation, but are not recommended unless minimum overhaul labor cost is of primary importance. When installing hand push fit sleeves, it may be necessary to select sleeves from more than one set to avoid too loose a fit. If sleeves fit too tightly to be pushed in by hand and maximum sleeve diameter and minimum cylinder diameter is within tolerance, either select another sleeve or chill the sleeve in dry ice or in deep freeze prior to installation. Hand push fit cylinder sleeves should never be installed in a diesel engine.

Regular press fit type cylinder sleeves are available for service in either semi-finished bore which requires honing or with finished bore in which the pistons and rings may be installed without honing. Optimum piston fit may be obtained with the semi-finished cylinder sleeves, following the honing procedure outlined in paragraph 104.

Reboring and installation of oversize pistons and rings, rather than resleeving, is the recommended overhaul procedure for all engines, sleeved and unsleeved; except where hand push fit sleeves have been installed.

Nominal standard cylinder bores are given below for the three engine sizes; however, pistons must be individually fit using a pull scale and feeler ribbon of the specified thickness, as outlined in paragraph 104.

Standard cylinder sizes are as follows:

134 cubic inch engine.. 3.4368-3.4388

144 cubic inch engine.. 3.5598-3.5618

172 cubic inch engine.. 3.8998-3.9018

Oversizes of 0.020, 0.030 and 0.040 are available for the pistons and rings of all models. In addition, pistons of 0.001 oversize for non-diesel engines and 0.002 oversize for diesel engines are available for service. Use standard size ring sets with 0.001 and 0.002 oversize pistons. Cylinders should be rebored if badly scored; if pistons cannot be fitted as outlined in paragraph 104; if out-of-round exceeds 0.003 or if taper exceeds 0.008. When reboring cylinder walls, leave 0.0015 stock below the minimum finished diameter, for finish honing and selective fitting of pistons. After cylinders have been rebored, finish hone as outlined in paragraph 105.

Rebore sizes for fitting of service sleeves in unsleeved engine blocks are as follows:

134 Cu. In. Engine

Block bore diameter....3.650-3.651

Counterbore diameter...3.748-3.752

Counterbore depth......0.179-0.180

144 Cu. In. Engine

Block bore diameter....3.720-3.721

Counterbore diameter...3.826-3.828

Counterbore depth 0.179-0.180

172 Cu. In. Engine

Block bore diameter.....4.100-4.101

Counterbore diameter...4.248-4.252

Counterbore depth 0.177-0.181

104. FITTING PISTONS. Before checking piston fit, deglaze the cylinder wall using a hone or deglazing tool. Attach a ½-inch wide feeler ribbon of specified thickness to a piston pull scale. Invert the piston, keeping pin bore parallel to crankshaft; and insert piston and feeler ribbon (90° from pin bore) into cylinder, until edge of piston skirt is approximately ½-inch below top of cylinder. Slowly withdraw the feeler ribbon, using the pull scale, while

noting scale reading. Fit is correct when 5-10 pounds pull is required to withdraw feeler ribbon.

Recommended thickness of feeler ribbon is as follows:

Non-diesel Engines:

New piston in new bore.......0.0015

New piston in used bore.......0.002

Used piston in used bore.......0.003

Diesel Engines:

New piston in new bore.......0.005

New piston in used bore.......0.006

Used piston in used bore.......0.007

105. FINAL SIZING OF CYLINDERS. Use a rigid type hone and 220 to 280 grit stones. A drill with a speed of 250 to 450 rpm should be used to drive the hone. The stones must be used dry to obtain the desired cylinder sleeve finish. Cover the crankshaft with clean rags.

NOTE: The speed of the hone and rapidity of the stroke govern the crosshatch marks on the sleeve. The crosshatch mark should intersect at approximately 90 degrees for proper ring seating.

Operate the hone through the bore 10 or 12 complete strokes. Remove the hone, clean the sleeve with dry rags, and recheck the piston fit.

Repeat the above procedure until the specified piston-to-bore fit is obtained.

CAUTION: Do not use gasoline or kerosene to clean the cylinder walls after the honing operation. Solvents of this type will not remove the abrasive but will further imbed small abrasive parties into the pores of the cylinder.

CLEANING AFTER HONING. After the honing is completed, clean the cylinder block of all foreign material as follows:

Wipe or remove as much of the abrasive material as possible.

Swab each cylinder wall with clean SAE 10 engine oil at least twice.

Wipe the oil out of the cylinder with clean rags.

Wash the cylinder bores with hot soapy water. Continue washing cylinder bore until clean white cloth can be rubbed through bore without discoloration.

Flush water jackets to remove foreign material which might cause excessive wear to the water pump.

Remove the rag from the crankshaft and wash the crankshaft with soapy water.

Dry the cylinder block thoroughly, using compressed air.

PISTON PINS

Non-Diesel Models

106. The 0.9120-0.9123 diameter floating type piston pins are retained in the piston pin bosses by snap rings and are available in standard size only. Piston pin should have a clearance of 0.0002-0.0008 in rod and 0.0001-0.0005 in piston. Bushings must be final sized and the oil hole drilled after installation. When assembling, oil squirt hole in rod must face toward camshaft side of engine and dimple in top face of piston must face toward front of engine.

Diesel Models

107. The floating type piston pins are retained in the piston bosses by snap rings and are available in standard sizes only. The piston pins used in the 144 cu. in. engine have a diameter of 1.1240-1.1243 while the 172 cu. in. engine uses a pin of 1.2490-1.2493 diameter. Piston pin should have a clearance of 0.0004-0.0007 in connecting rod bushing and 0.0001-0.0003 in piston. Bushings must be final sized after installation. When assembled, oil squirt hole in rod must face toward camshaft side of engine and "FRONT" mark in top face of piston must face toward front of engine.

CONNECTING RODS AND BEARINGS

All Models

108. Connecting rod bearings are of the non-adjustable, slip-in precision type, renewable from below after removing the oil pan and connecting rod bearing caps. When installing new bearing shells, make certain that the bearing shell projections engage the milled slot in connecting rod and bearing cap and that cylinder numbers on the rod and cap are in register. If rod and cap are not numbered, be sure bearing shell projections in both rod and cap are to same side. Bearing inserts are the same for all cylinders and are available in standard sizes as well as undersizes of 0.001, 0.002, 0.003, 0.010, 0.020, 0.030 and 0.040 for non-diesel engines and undersizes of 0.001, 0.002 and 0.003 for diesel engines.

Check the crankshaft crankpins and the rod bearing inserts against the values which follow:

Crankpin diameter......2.2978-2.2988
    Regrind if out-of-round.....0.001
    Regrind if tapered..........0.001
    NOTE: Ford Motor Co. states that diesel crankshafts should not be reground to fit undersize bearing inserts.

Bearing running
    clearance 0.0007-0.0027
    Renew bearing if clearance exceeds 0.003
Rod side play..........0.003 -0.009

NOTE: Bearing clearance of 0.0007-0.0027 is for copper-lead type inserts currently used in all engines. If early type babbitt inserts are used in 134 cu. in. engines, desired bearing clearance is 0.0006-0.0021.

CAUTION: Never use babbitt type inserts in diesel engines or in 172 cu. in. non-diesel engines.

Connecting rod center to center length is 6.258-6.262.

When reassembling, tighten the connecting rod nuts to a torque of 45-50 Ft.-Lbs. Install new pal nuts and tighten them to a torque of 3-4 Ft.-Lbs. (or finger tight plus ⅓ turn).

CRANKSHAFT AND MAIN BEARINGS

NOTE: A few non-diesel engines have been shipped with diesel (forged steel) crankshafts instead of cast iron crankshafts as original equipment. On these engines, use diesel connecting rod bearing liners.

All Models

109. Crankshaft is supported in three main bearings of the non-adjustable, slip-in precision type, renewable from below after removing the oil pan, oil pump pipes and main bearing caps. Bearing inserts are available in standard size, as well as undersizes of 0.001, 0.002, 0.003, 0.010, 0.020, 0.030 and 0.040 for non-diesel engines, and 0.001, 0.002 and 0.003 for diesel engines. Normal crankshaft end play of 0.002-0.006 is controlled by the flanged center main bearing inserts. Renew inserts if end clearance exceeds 0.010.

NOTE: Ford Motor Co. states that diesel engine crankshafts should not be reground for undersize bearings.

To remove the crankshaft, first remove engine as in paragraph 82, 83 or 84; then, remove the clutch, flywheel, timing gear cover, oil pan, oil pump, engine end plate and rod and main bearing caps. Lift crankshaft from cylinder block.

Check the crankshaft and main bearing inserts against the values which follow:

Non-Diesel (Cast Iron) Crankshaft

Crankpin diameter 2.2978-2.2988
    Regrind if out-of-round......0.001

Regrind if tapered..........0.001
Main journal diameter..2.4974-2.4988
    Regrind if out-of-round......0.001
    Regrind if tapered..........0.001

Main bearing running
    clearance 0.0005-0.0027
    Renew bearing if clearance exceeds 0.003

Crankshaft end play....0.002 -0.006
    Renew bearing if end play exceeds 0.010
Journal radii ....0.08 -0.10
Main bearing
    bolt torque ....95-105 Ft.-Lbs.

Diesel (Forged Steel) Crankshaft

Crankpin diameter......2.2978-2.2988
    Renew crankshaft if out-of-round 0.001
    Renew crankshaft if tapered 0.001

Main journal diameter..2.4968-2.4977
    Renew crankshaft if out-of-round 0.001
    Renew crankshaft if tapered 0.001

Main bearing running
    clearance 0.0019-0.0037

Renew bearing if clearance exceeds 0.005

Crankshaft end play....0.002 -0.006

Renew bearing if end play exceeds 0.010

Journal radii ....0.110 -0.120

Main bearing bolt
    torque ....110-115 Ft.-Lbs.

Crankshaft end play is controlled by the flanges on the center main bearing inserts. When installing the center main bearing cap, tighten the cap screws finger tight; then, bump the crankshaft back and forth to align the bearing thrust surfaces before tightening the cap screws.

The crankshaft rear oil seals are located in the cylinder block and rear main bearing cap. Renew the seals as outlined in paragraph 111. Note: The flat parting surfaces of the rear main bearing cap and cylinder block are used for sealing. Before installing the rear main bearing cap, be sure these surfaces are clean and smooth. If rope type rear crankshaft seal is used, be sure the ends are cut cleanly and no fiber strands extend from the sealing grooves. Also, some mechanics apply a very thin coat of non-hardening gasket sealer on parting surfaces before installing rear main bearing cap.

NOTE: Flat washers are used under the main bearing retaining cap screw heads on diesel engines.

CRANKSHAFT OIL SEALS

110. FRONT OIL SEAL. On all models, the crankshaft front oil seal is located in the timing gear cover and can be renewed after removing the cover as outlined in paragraph 96. Install seal with lip towards inside of cover. The seal lip seats against the hub of the crankshaft pulley; renew the pulley hub if groove is worn at sealing surface. Note: On early production non-diesel models with separate governor housing, it is not necessary to remove the timing gear cover; front oil seal may be renewed after removing governor housing from timing gear cover.

111. REAR OIL SEAL. Two different types of rear crankshaft oil seal are available.

A two-piece neoprene lip type seal, which can be installed with the crankshaft in place, is provided with engine lower service gasket kits. To install this type seal, remove old seal from groove in cylinder block and rear main bearing cap. Then place one half of the seal against the crankshaft with sealing lip forward and rotate it into the groove in cylinder block. Place the other half of the seal against the crankshaft with sealing lip forward and push one end of the seal into the groove in cylinder block until the parting line of the seal halves is at about 45 degrees angle to the parting line of cylinder block and main bearing cap.

An optional rope type seal, Ford part No. EBU-6701-A, can be installed only when the crankshaft is removed and, for proper sealing, should only be installed using a special oil seal forming tool (Nuday tool No. 6701-A). Install the rope type seal as follows: Soak the seal in oil at least ½ hour prior to installation. Work the seals into the grooves in cylinder block and rear main bearing cap so that both ends of each half of the seal protrude evenly. Using the seal forming tool with flange on tool inserted in crankshaft oil slinger groove, drive on the tool handle with a hammer until the seals are seated (flange on tool bottoms in oil slinger groove), then, while holding tool firmly against the seal, cut off the seal ends flush with block and main bearing cap with sharp knife. The seal forming tool has plastic inserts to cut against.

In addition to the shaft seals, two wedge type seals are inserted at the sides of the rear main bearing cap. After the rear main bearing cap is installed, dip the wedge seals in oil and quickly insert them in the grooves at each side of main bearing cap. The seals are slightly longer than the groove in the main bearing cap; tap against end of seal with hammer until seal is flush with cylinder block and bearing cap. CAUTION: Do not cut any material from the ends of the wedge seals. Note: At tractor Serial No. 138247 of 1958-62 production, size of grooves in sides of rear main bearing cap was decreased to provide a more positive seal; 0.010 oversize seals are available for tractors prior to this serial number.

ENGINE BALANCER

Diesel Models

112. A Lanchester type engine balancer was used on the 172 cubic inch diesel engine starting with serial number 102357 of 1958-62 production. The balancer became standard equipment on 144 cubic inch engines starting with serial number 128685 of 1958-62 production. A service balancer is available which may be installed on any Ford four cylinder diesel engine prior to these serial numbers.

The balancer is driven at twice engine speed by a gear machined on the crankshaft and located immediately behind the center main bearing. Installation of the balancer on tractors not so equipped requires changing crankshaft and oil pump lines.

When installing the balancer, turn crankshaft until No. 1 piston is at top center position, and align the scribed timing marks on crankshaft gear and balancer gear. Shift the balancer frame sideways if necessary until the backlash between crankshaft gear and balancer drive gear is 0.005-0.012. On factory installed balancers, tighten the large, hollow retaining cap

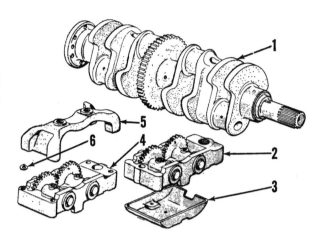

Fig. FO78 — Crankshaft and balancer assembly used on diesel engines.

1. Crankshaft
2. Balancer
3. Baffle
4. Balancer (service)
5. Adapter (service)
6. Shim

screw to a torque of 70-80 Ft.-Lbs. and the smaller cap screw to a torque of 60-70 Ft.-Lbs. On service installed balancers, tighten the four retaining cap screws to a torque of 28 Ft.-Lbs. NOTE: Washer type shims (0.003) are provided for the service balancer to be installed between adaptor and balancer frame. Shim thickness should be equal at all four attaching points. Shims affect the clearance between balancer frame and cylinder block. Add or remove shims as required to obtain the recomended clearance.

When installing service balancer on tractors not so equipped, the cap screw flats on the center main bearing cap must be milled to a thickness of 2.031-2.033, with the thickness of both sides being exactly equal. Install the adaptor using the special main bearing cap screws provided. Tighten the cap screws to a torque of 110-115 Ft.-Lbs. Lay a straight edge across the oil pan mounting flanges on cylinder block and balancer mounting surfaces of adaptor. The four surfaces should be parallel, and balancer mounting surfaces should be approximately flush with oil pan attaching flanges. Measure the clearance at adaptor flange or oil pan flange; then add the 0.003 washer shims as required to bring the balancer mounting points 0.001-0.003 beyond the oil pan mounting flange. Remove screen from oil pan drain plug and saw plug off at screen retaining groove. Install new oil lines, oil pan and modified drain plug.

113. OVERHAUL. Before attempting to disassemble the balancer, inspect to determine the method of retaining the counterweight gears to shaft. In factory assembly, the counterweight is retained by a staked steel ball which fits into shaft indentation. In service assembly, the counterweight is retained by a special $\frac{7}{16}$-20, cone point set screw. Disassemble the balancer as follows:

If counterweight gear is retained by a staked ball, remove the shaft by using a suitable press. Counterweight gears may be reused, if serviceable, by driving out the steel ball with a hammer and punch; then enlarging the ball hole with a 25/64-inch drill. Thread the drilled hole using a $\frac{7}{16}$-NF thread tap, and install the special $\frac{7}{16}$-20 cone point set screw. Use a new gear shaft when reassembling.

If counterweight gear is retained by a set screw, remove the set screw; then tap the shaft from balancer assembly.

Remove the four needle roller bearings from balancer frame by inserting a $\frac{5}{8}$-inch rod or shaft through upper bearing; then using a $\frac{15}{16}$-inch step plate to press the lower bearing toward the OUTSIDE of housing. Install new bearing in same manner; pressing the bearing in from INSIDE the housing to keep from distorting the balancer frame. Bearing is correctly positioned when end of bearing is flush with outside of frame bore.

NOTE: When installing bearings, press on the numbered or lettered end of bearing, to keep from distorting the outer cage.

When assembling the balancer, make sure that timing marks on the two balancer counterweight gears are aligned; and that the retaining set screw enters notch in gear shaft. Tighten the set screws to a torque of 20 Ft.-Lbs. and stake in place.

FLYWHEEL

All Models

114. The flywheel can be removed after detaching the engine from the transmission and removing the clutch. Flywheel can be reinstalled in one position only. Tighten the flywheel retaining bolts to a torque of 75-85 Ft.-Lbs.

Starter ring gear is installed from rear face of flywheel and thus can be renewed without removing flywheel from crankshaft. To facilitate installation of the gear, heat same uniformly to 400 degrees F. using an oxyacetylene torch having a No. 2 or smaller tip. Adjust torch for a rich acetylene mixture. Apply heat to inside of ring gear only; do not touch the gear teeth with the flame. Stop applying heat immediately when gear is just hot enough to melt a 400° F. temperature indicating crayon and quickly install the gear against

shoulder on flywheel. Be sure gear is properly installed; then quench the gear with water to cool it rapidly. Caution: Do not heat gear above 450° F.

OIL PAN (SUMP)
All Models

115. **REMOVE AND REINSTALL.** Oil pan can be unbolted and removed on all models without removing other components. Pan is cast iron on Offset models and stamped steel on other models.

When reinstalling pan, use a thin film of heavy grease to stick new gasket to gasket surface of cylinder block. Tighten the pan retaining cap screws to 20-25 Ft.-Lbs. torque on Offset models and to 12-15 Ft-Lbs. torque on other models.

OIL PUMP AND RELIEF VALVE

A gear type oil pump such as that shown in Fig. FO82 was used in all early tractors and through part of the 1961 production. Starting in 1961, a gyrotor type oil pump such as that shown in Fig. FO79 is used. The new pump may be installed on tractors equipped with the gear-type pump, provided the inlet screen, pressure tube and attaching parts are also changed.

G-Rotor Type

116. Refer to Fig. FO79. To disassemble the pump, remove the cover (8) and withdraw the inner rotor (6) and outer rotor (7). Examine the parts for wear or scoring. Relief valve spring (3) should test 9.8 lbs. when compressed to a height of 1.56 inches.

To check the rotor clearances, reassemble rotors in housing, making sure the correlation marks on outer and inner rotors are on the same side. Measure the clearance between points of inner rotor and mating surface of outer rotor as shown in Fig. FO81. Clearance should be 0.003-0.008 with rotors in any position. Lay a straight edge across gasket surface of pump body (See Fig. FO80) and measure clearance between straight edge and lower surface of rotors. The clearance should not exceed 0.004. Renew any parts which are scored, worn or otherwise damaged. Assemble by reversing the disassembly procedure. Tighten the oil pump retaining cap screws to a torque of 30-35 Ft.-Lbs., and the cover cap screws to 10-15 Ft.-Lbs.

Gear Type

117. To remove the pump, remove the oil pan, oil filter pipe (14—Fig.

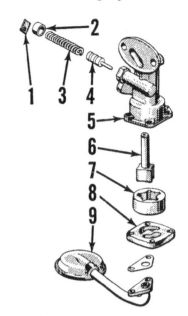

Fig. FO79 — Exploded view of G-Rotor type oil pump used on late models.

1. Retainer	6. Shaft & rotor
2. Plug	7. Outer rotor
3. Valve spring	8. Cover
4. Relief valve	9. Inlet tube & screen
5. Body	

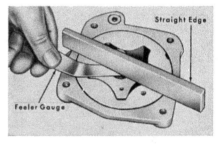

Fig. FO80 — Use straight edge when checking rotor end clearance. Clearance should not exceed 0.004.

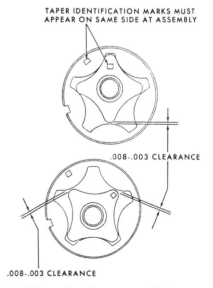

Fig. FO81 — Install rotor assembly in pump body and check clearances as shown.

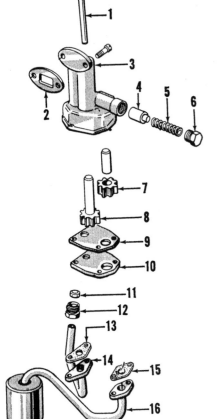

Fig. FO82 — Exploded view of early gear type engine oil pump assembly.

1. Hex drive shaft	9. Gasket
2. Gasket	10. Cover plate
3. Pump body	11. Rubber seal
4. Relief valve	12. Packing nut
5. Spring	13. Gasket
6. Cap	14. Oil pressure tube
7. Driven gear	15. Gasket
8. Drive gear & shaft	16. Oil inlet tube

FO82) and the Allen head bolts. Withdraw the pump and inlet pipe assembly from engine. Disassemble the pump and check the component parts against the following values:

Drive shaft clearance
   in pump body........0.0015-0.0029

Driven gear to
   pin clearance 0.001-0.002

Gear side clearance
   in pump body..........0.005 Max.

Relief valve spring
   tension ......9.8 lbs. @ 1.56 inches

Relief valve opening
   pressure .....45-50 psi @ 1400 rpm

Renew the oil pump cover plate (10) if machined surface of same shows wear. If necessary to renew the pump body, the gear type pump must be discarded and a new G-rotor type pump installed. See note preceding paragraph 116.

When reassembling, tighten the cover cap screws to a torque of 15-18 Ft.-Lbs. and the pump mounting bolts to a torque of 30-35 Ft.-Lbs.

CARBURETOR
(Except LP-Gas)

118. Marvel - Schebler carburetors are used. Clockwise rotation of the air controlling idle adjusting needle will enrich the mixture, whereas clockwise rotation of the main adjusting needle will lean the mixture. Initial setting for the main adjusting needle is 1-1¼ turns open for series 800 and 900, 1 turn open for series 600 and 700 and 1¼ turns open for later models. Adjust the throttle stop screw to obtain a slow idle speed of 450-475 rpm. Float setting is ¼-inch from nearest face of cloat to gasket on bowl cover.

Model numbers are as follows:

Model TSX580
Repair kit 286-1095
Gasket set 16-675
Inlet needle & seat.........233-536

Model TSX593
Repair kit 286-1163
Gasket set 16-634
Inlet needle & seat.........233-536

Model TSX662
Repair kit 286-1183
Gasket set 16-654
Inlet needle & seat.........233-595

Model TSX692
Repair kit 286-1185
Gasket set 16-634
Inlet needle & seat.........233-595

Model TSX706
Repair kit 286-1163
Gasket set 16-634
Inlet needle & seat.........233-595

Model TSX765
Repair kit 286-1263
Gasket set 16-634
Inlet needle & seat.........233-595

Model TSX769
Repair kit 286-1264
Gasket set 16-654
Inlet needle & seat.........233-595

Model TSX813
Repair kit 286-1345
Gasket set 16-654
Inlet needle & seat.........233-595

LP-GAS SYSTEM

Both Ford non-diesel tractor engines are available with a factory installed LP-Gas system. A flow diagram through the various system components is shown in Fig. FO85. Fuel tank capacity is 25 U. S. gallons, but tank should NEVER be filled more than 80% full (20 U. S. gallons). The remaining 20% allows for expansion of the fuel due to a possible rise in temperature.

CAUTION: LP-Gas expands readily with any decided increase in temperature. If tractor must be taken into a warm shop to be worked on during extremely cold weather, make certain that fuel tank is as near empty as possible. LP-Gas tractors should never be worked on or stored in an unventilated space.

PRINCIPLES OF OPERATION

119. Refer to Fig. FO85. With the fuel tank liquid withdrawal valve open, liquid fuel under pressure flows through the filter assembly to the inlet valve of the vaporizer (A). Spring pressure unseats the inlet valve (B) and the fuel flows into the vaporizing chamber (C) where it expands into a vapor. Spring action on the inlet valve, combined with gas pressure acting on the diaphragm at (D), opens and closes the inlet valve (B) until a constant regulated gage pressure of 10 psi is maintained in chamber (C).

With the engine turning over, a vacuum condition (slightly less than atmospheric pressure) is set up in the chamber (E) of the regulator. The balance line applies atmospheric pressure to the outside of the regulator diaphragms (G). This difference in pressure between the outside and inside of the diaphragms causes the two regulator diaphragm assemblies to draw inward against the leaf spring portion of the valve block assembly (F), lifting the valve block from its seat in the inlet orifice. Vaporized fuel then passes from the vaporizer outlet and into the chamber (E) at a pressure which is less than atmospheric. When the engine is stopped, the pressure in the chamber (E) rises to atmospheric and the valve block assembly (F) closes under the force of coil spring (I).

The system of passages in the carburetor (J) sets up a condition of

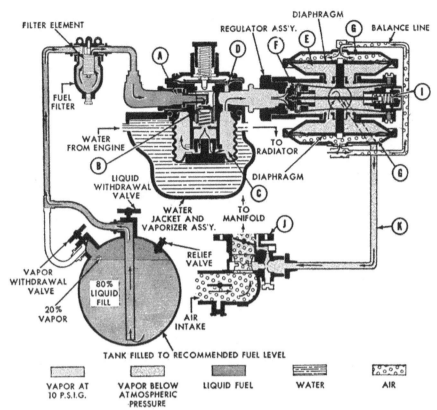

VAPOR AT 10 P.S.I.G.   VAPOR BELOW ATMOSPHERIC PRESSURE   LIQUID FUEL   WATER   AIR

Fig. FO85 — Schematic view showing the fuel flow through the LP-Gas system. The LP-Gas system is of the liquid withdrawal type.

vacuum in the tube (K) and chamber (E) that results in the proper flow of fuel to the engine.

The carburetor choke butterfly must be temporarily closed to obtain a richer mixture for starting purposes.

TROUBLE-SHOOTING

120. The following trouble-shooting paragraphs list troubles which can be attributed directly to the fuel system; however, many of the troubles can be caused by derangement of other parts such as valves, battery, spark plugs, distributor, coil, etc.

The procedure for remedying many of the causes of trouble is evident. The following paragraphs will list the most likely causes of trouble, but only the remedies which are not evident.

ENGINE WILL NOT START. Could be caused by:

a. Choke not closing when control is pulled out. Choke must be closed momentarily when starting a cold engine to start the vaporized fuel flowing. Do not hold choke closed for more than two or three seconds while cranking or engine may become flooded. Clean out a flooded engine by cranking several revolutions with choke and throttle wide open.

b. Improperly blended fuel. An excess of Butane will cause hard or no starting. Since Butane is liquid at temperatures below 32 degrees F., there will be an insufficient quantity of vapor for starting purposes.

c. Excess flow valves in the fuel tank withdrawal valves closed. If the liquid or vapor withdrawal valves are opened too fast, the excess flow valves will shut-off the fuel flow automatically. Close the withdrawal valves to reset the excess flow valves; then, open the withdrawal valves slowly.

d. Improperly adjusted regulator.

INSUFFICIENT ENGINE POWER. Could be caused by:

a. Improperly adjusted carburetor.

b. Clogged or restricted air filter.

c. Leaking carburetor or manifold connections.

d. Loose fuel line connections.

e. Improperly adjusted regulator.

f. Faulty ignition timing.

g. Improperly adjusted spark plug gap. Gap should be 0.028-0.033.

ENGINE FALTERS OR DIES ABOVE IDLE SPEEDS. Could be caused by:

a. Improperly adjusted carburetor.

b. Clogged or restricted air filter.

c. Leaking carburetor or manifold connections.

d. Loose fuel line connections. Check with soapy water.

e. Improperly adjusted regulator.

f. Improperly adjusted spark plug gap. Gap should be 0.028-0.033.

g. Clogged fuel filter element.

CARBURETOR

121. **ADJUSTMENT.** Clockwise rotation of the idle and main adjustment screws leans the mixture. Refer to Fig. FO86. Initial idle needle setting is one turn open. Initial main adjustment needle setting is 5½ turns open. To adjust the carburetor, start engine and bring to operating temperature then turn the throttle stop screw either way as required to provide an engine slow idle speed of 450-475 rpm. Turn the idle adjusting needle clockwise until engine begins to "stall"; then, turn the screw counterclockwise until engine runs smoothly. To adjust the main adjusting needle, bring engine up to operating temperature, remove any three spark plug wires, open the throttle to full throttle setting and adjust the main needle until highest operating rpm is obtained. Turn the needle clockwise until engine speed just begins to drop, then turn needle counter-clockwise just enough to bring the engine speed back to top rpm.

Recheck low idle rpm.

122. **R&R AND OVERHAUL.** Before removing any of the fuel system components, close both withdrawal valves; and start and run the engine until all fuel is consumed and engine stops.

Two types of Zenith carburetors have been used. On early models, a secondary regulator is located on the vaporizer assembly and fuel enters the carburetor at approximately atmospheric pressure. Starting with serial number 131530 of 1958-62 production, a pressure regulating carburetor is used, incorporating a secondary regulator. On these models, fuel enters the carburetor at the 10 psi. vaporizer outlet pressure. Removal procedures are similar for the two units, but method of overhaul will differ.

Fig. FO86 — Installation view of early LP-Gas carburetor, showing the mixture adjusting needles and the throttle stop screw.

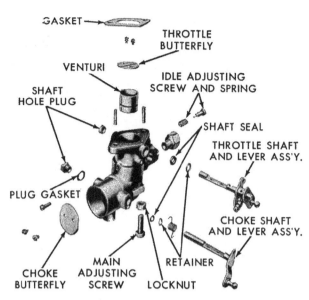

Fig. FO87 — Exploded view of early LP-Gas carburetor. The carburetor is provided with two mixture adjustments plus a throttle stop screw.

Early Models: Except for the removal of air-borne dirt and resealing of throttle and choke shafts, overhaul is not generally required. Removal of venturi is rarely necessary. If venturi is to be removed, first remove throttle valve and shaft, and invert carburetor body; then remove the venturi locking screw located above choke valve. Install venturi so that step in counterbore faces upward. Remove burrs in choke and throttle shaft bores caused by previous staking; install new seals and retainers, and stake new seal retainers 1/8-inch below outside of body bores.

Late Models: Mixing chamber components are similar to those used on early models, and overhaul procedures are similar. The carburetor incorporates a dual diaphragm, pressure regulator which reduces the vaporizer fuel pressure to approximately carburetor inlet pressure for efficient and economical operation. Carburetor is completely sealed.

Inlet valve seat (10—Fig. FO88) is locked in position by nylon plug (15) and set screw (16). The position of the seat controls the opening position of diaphragm lever (17). When assembling or adjusting the carburetor, loosen the plug (16) and thread the valve seat (10) in or out until valve lever (17) just contacts Step 2 of Zenith Adjustment Gage C161-194. If a gage is not available, adjust the valve seat until distance (A—Fig. FO89) measures 3/16-inch.

Inspect the diaphragm (19 & 21—

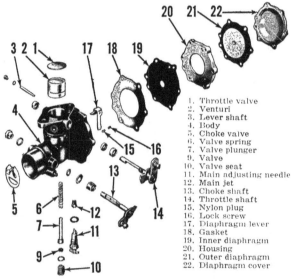

1. Throttle valve
2. Venturi
3. Lever shaft
4. Body
5. Choke valve
6. Valve spring
7. Valve plunger
9. Valve
10. Valve seat
11. Main adjusting needle
12. Main jet
13. Choke shaft
14. Throttle shaft
15. Nylon plug
16. Lock screw
17. Diaphragm lever
18. Gasket
19. Inner diaphragm
20. Housing
21. Outer diaphragm
22. Diaphragm cover

Fig. FO88 — Exploded view of pressure regulating LP-Gas carburetor used on late models.

Fig. FO89 — Schematic view of regulating valve assembly.

A. Adjustment
L. Lever
S. Valve seat
V. Valve

Fig. FO88) for pin holes or cracks, and install so that cupped spacer on inner diaphragm (19) faces toward the steel center plate of outer diaphragm (21). Make sure the passage holes in diaphragms, housings and gasket are aligned during assembly.

FUEL FILTER

123. To renew or clean the filter element, be sure that all fuel is burned out of the lines, regulator and carburetor by closing the tank withdrawal valve and allowing engine to run until it stops. Turn the filter cover counter-clockwise and remove same. Using caution not to damage the metal discs, remove filter element. Discs must be flat for proper filtering action. Renew the filter element unit if there is any evidence of damage or if discs are bent. A dirty element which is otherwise in good condition can be cleaned by rinsing same in a suitable solvent and directing compressed air through length of element. Disc damage can result if air stream is directed to the sides.

When reassembling, install the element just finger tight, install filter cover and tighten same securely.

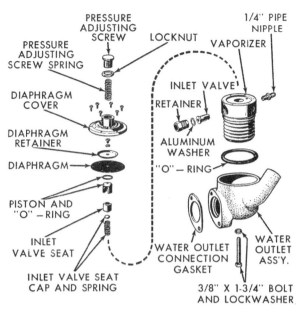

Fig. F090—Exploded view of early LP-Gas vaporizer.

Fig. FO92 — Using an air hose and pressure gage to adjust the LP-Gas vaporizer working pressure to 10 psi.

Fig. FO91 — Using a ¼-20 screw to remove the vaporizer inlet valve. Pressure adjusting screw must be turned completely down during this operation.

REGULATOR AND VAPORIZER

Early Models

124. **R&R AND OVERHAUL.** Before disconnecting any lines, be sure that all fuel is burned out of the lines, regulator and carburetor by closing the tank withdrawal valve and allowing engine to run until it stops. Remove hood, drain cooling system and disconnect the upper radiator hose. Disconnect the lines and hose from the regulator, back off the cap screw at underside of water jacket two or three turns, strike the screw sharply

with a soft-nosed hammer to free the vaporizer assembly.

125. **VAPORIZER OVERHAUL.** Remove the inlet valve seat retainer (Fig. FO90) and aluminum washer. Loosen the pressure adjusting screw locknut and turn the adjusting screw completely down to depress the inlet valve seat. Then screw a ¼-20 cap screw into the inlet valve and remove the valve as shown in Fig. FO-91. Note: The inlet valve seat must be fully depressed to prevent damage to seat and valve parts.

Remove the pressure adjusting

screw and spring, then unscrew and remove the diaphragm cover. Lift the rubber diaphragm and piston assembly from the bore; then remove and discard the rubber diaphragm. Remove the inlet valve seat, spring and spring cap. Inspect and renew any damaged or worn parts.

When reassembling, install a new diaphragm on the piston but do not tighten the diaphragm retaining screw at this time. Temporarily install the inlet valve in its bore. Notice that slot of valve is positioned by dowel pin in its bore. Position the assembled diaphragm and piston into its bore so that the piston straddles the inlet valve without rubbing against it. Align the six diaphragm screw holes with the screw holes in the vaporizer; then, while preventing the diaphragm or piston from turning, tighten the screw retaining the diaphragm to the piston. After proper alignment has been obtained, remove the diaphragm and piston assembly and the inlet valve from their respective bores.

Install the assembled inlet valve seat, cap and spring into the vaporizer. Install a new well lubricated "O" ring on piston and install the piston and diaphragm assembly so that yoke in piston registers with the inlet valve bore and screw holes in diaphragm line up with screw holes in vaporizer. Install top cover, adjusting screw spring, adjusting screw and lock nut. Turn the adjusting screw all the way in to depress the inlet valve seat; then, using a new "O" ring on the inlet valve, insert the valve and remove the ¼-20 cap screw. Install a new aluminum washer, then install and tighten the inlet valve seat retainer. Install new "O" ring on vaporizer.

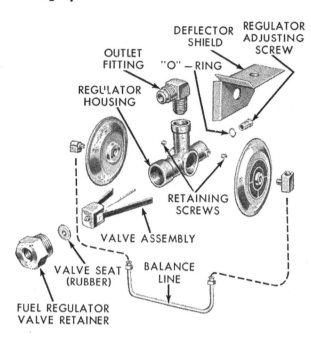

OUTLET FITTING · DEFLECTOR SHIELD · REGULATOR ADJUSTING SCREW · "O" – RING · REGULATOR HOUSING · RETAINING SCREWS · VALVE ASSEMBLY · VALVE SEAT (RUBBER) · BALANCE LINE · FUEL REGULATOR VALVE RETAINER

Fig. FO93 — Exploded view of the LP-Gas system pressure regulator.

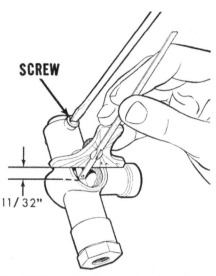

SCREW

11/32"

Fig. FO96 — Regulator leaf springs should be depressed 11/32-inch with a depth gage before tightening the retainer screws.

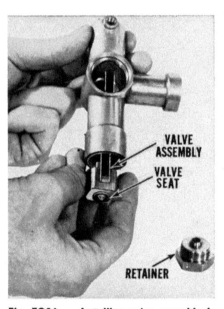

VALVE ASSEMBLY · VALVE SEAT · RETAINER

Fig. FO94 — Installing valve assembly in the regulator housing. Leaf springs must enter slots at opposite end of bore.

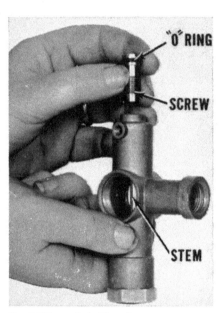

"O" RING · SCREW · STEM

Fig. FO95 — Installing the regulator pressure adjusting screw. Stem of valve must enter the hole in the screw.

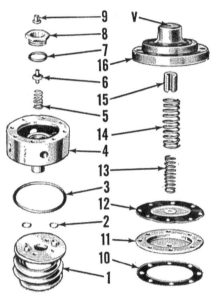

Fig. FO97 — Exploded view of late model vaporizer.

1. Coil & plate	10. Gasket
2. "O" ring	11. Baffle
3. "O" ring	12. Diaphragm
4. Body	13. Inner spring
5. Valve spring	14. Outer spring
6. Valve	15. Spacer
7. "O" ring	16. Cover
8. Valve seat	V. Vent hole
9. Follower	

Before installing the vaporizer, adjust the unit as follows: Clamp the assembly in a soft jawed vise and connect an air hose which will supply approximately 75 psi to the inlet connection. Connect a 0 to 30 pressure gage to the outlet connection as shown in Fig. FO92. Back out the pressure adjusting screw until only one or two threads are holding the screw in the cover and apply air pressure. Turn the screw in slowly until the gage shows an outlet pressure of 10 psi. Then loosen the gage to bleed off some of the air, retighten the gage,

recheck the gage reading and readjust if necessary. If it is impossible to obtain the desired 10 psi gage reading, recheck the vaporizer for improper assembly or improper parts. Tighten the adjusting screw lock nut.

126. REGULATOR OVERHAUL. Refer to Fig. FO93. Disassemble and reassemble the regulator in the exact following sequence or internal parts of the regulator may be damaged. Remove the balance line and both connectors, then remove both of the diaphragm assemblies by turning them

counter-clockwise. Remove the two valve stem spring retaining screws and the regulator adjusting screw from the regulator housing. Remove the regulator valve retainer and withdraw the valve assembly from the housing bore. Remove the rubber seat from valve assembly. Inspect all parts and renew any which are damaged or worn. If leaf springs have taken a definite inward set, renew the valve assembly. Install a new rubber seat on valve assembly and insert the assembly into housing, making certain that the two leaf springs enter their respective locating slots at opposite end of bore. Refer to Fig. FO-94. Now, while holding the housing in a position to prevent the valve assembly from falling out, install and tighten the fuel regulator valve retainer. Install a new well lubricated "O" ring on the pressure adjusting screw and install the adjusting screw as shown in Fig. FO95. Make certain, however, that stem of valve assembly enters the hole of the adjusting screw before turning the screw inward until it seats firmly. Note: Later production regulator is not equipped with adjusting screw.

Inspect the nylon seats of the two retainer screws and renew the screws if seats are extremely flattened. Position the regulator assembly with the valve retainer downward so the installed valve assembly will be resting on it. Use a depth gage as shown in Fig. FO96, depress each of the leaf springs 11/32-inch as shown and while holding the springs in the depressed position, tighten the retainer screws. Install the two diaphragm assemblies to the housing and tighten them hand tight. Use a non-hardening pipe sealer on all connections to be sure they are air tight and install the remaining parts.

Late Models

127. The vaporizer assembly is non-adjustable. A complete diagnosis of vaporizer condition can be made without disassembly of the unit, and a test is recommended before overhaul is attempted. To make the test, proceed as follows:

Connect the vaporizer inlet to a source of compressed air and plug the outlet port. Completely immerse the vaporizer in water and inspect for air leaks. Note especially, the areas around vaporizer coil and mounting plate (1—Fig. FO97) and around diaphragm cover (16). Air bubbles emerging from vent hole (V) in top of cover indicate a leaking diaphragm.

To check the vaporizer inlet valve and seat, install a 0-300 psi pressure gage in vaporizer outlet port and connect inlet to air pressure. The gage reading should rise to 9-11 psi and hold steady. If pressure continues to rise with outlet plugged, a leaking fuel valve, or valve seat "O" ring is indicated, and vaporizer should be overhauled.

To disassemble the vaporizer, remove four alternate screws from diaphragm cover (16) and install alignment studs (Zenith Part No. C161-195). Apply thumb pressure to top of diaphragm cover while remaining screws are removed.

When reassembling, install fuel valve (6) with long stem toward diaphragm. Make sure all screw holes are aligned in gasket (10), baffle (11), diaphragm (12) and cover (16). Tighten the retaining screws evenly, leaving the alignment studs installed, until cover is tight. Recheck for leaks after assembling by immersing in water or using a soap solution.

DIESEL FUEL SYSTEM

The diesel fuel system consists of three basic components: the fuel filters, injection pump and injection nozzles. When servicing any unit associated with the fuel system, the maintenance of absolute cleanliness is of utmost importance. Of equal importance is the avoidance of nicks or burrs on any of the working parts.

Probably the most important precaution that service personnel can impart to owners of diesel powered tractors, is to urge them to use an approved fuel that is absolutely clean and free from foreign material. Extra precaution should be taken to make certain that no water enters the fuel storage tanks.

QUICK CHECKS

All Diesel Models

128. If the engine does not run properly and the fuel system is suspected as the source of trouble, refer to the accompanying trouble-shooting chart and locate points which require further checking. Many of the chart items are self-explanatory; however, if the difficulty points to the fuel filters, injection nozzles and/or injection pump refer to the appropriate sections which follow:

FILTERS AND BLEEDING

All Diesel Models

129. **MAINTENANCE.** The fuel filter is fitted with a disposable type element which should be renewed every 400 hours or every six months. However, element should be renewed whenever fuel will not flow freely from open bleed screw.

To renew the element first shut off fuel, then open drain cock at bottom rear of filter and drain filter base. Turn cover counter-clockwise and remove cover and filter element. The filter element is spring loaded and will rise from filter base as the cover assembly is unscrewed. See Fig. FO-100. Remove the rubber grommet and spring from the filter center post and discard grommet and element gaskets.

Clean all parts paying particular attention to the filter base; then using a new rubber grommet, assemble spring and grommet on the center post of filter base. Use new gaskets at top and bottom of filter element and place the filter element over grommet on center post of filter base. Place cover and bolt assembly through hole in filter element and while slightly depressing element, start bolt threads. Tighten

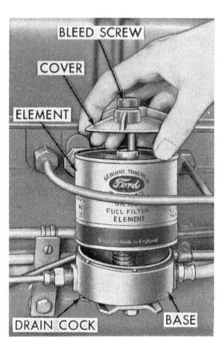

Fig. FO100 — Filter cover can be removed by turning counter-clockwise. Filter will lift from base as cover is unscrewed.

cover and bolt assembly finger tight. Open fuel shut-off and bleed fuel system as outlined in the following paragraph.

DIESEL SYSTEM TROUBLE-SHOOTING CHART

	Lack of Fuel	Engine Surging or Rough	Cylinders Uneven	Engine Smokes or Knocks	Injection Pump Does Not Shut Off	Engine Dies at Low Speed	Loss of Power
Defective Speed Control Linkage	★				★		★
Air in Fuel System	★			★		★	
Clogged Filter	★					★	★
Fuel Lines Leaking or Clogged	★	★	★			★	★
Friction in Injection Pump		★					★
Inferior or Contaminated Fuel		★		★		★	★
Faulty Injection Pump Timing		★		★		★	★
Defective Nozzle or Injector		★	★	★			★
Faulty Governor and/or Linkage Adjustment		★		★	★		★
Sticking Plunger		★				★	★
Faulty Primary Pump	★						★
Faulty Distribution of Fuel	★	★	★	★			★
Injection Pump Not Turning	★						
Friction in Governor		★					

130. **BLEEDING.** Open fuel tank shut-off valve and drain cock at bottom rear of filter base. Close the drain cock on filter base when fuel flows out. Open the air bleed screw located on the top of the fuel filter approximately two turns and observe the fuel flow. When the air bubbles disappear and the fuel flows in a solid stream, close the bleed screw. CAUTION: Never open the air bleed screw while engine is running. To do so will introduce air into fuel system. Remove the pump leak-off line at the governor cover and crank the engine until fuel flows in a solid stream from the fitting on pump.

If tractor does not start readily, loosen the injection line connections at the injectors while cranking engine, then retighten when fuel flows from the connection.

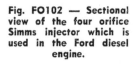

Fig. FO101 — To completely test an injector nozzle requires the use of a nozzle tester as shown.

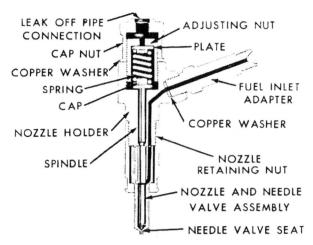

Fig. FO103 — With nozzle pressure brought to within 150 psi of opening pressure and held for one minute, the fuel stain should not exceed a ½-inch diameter when clean blotting paper is applied to nozzle tip.

INJECTION NOZZLES

All Diesel Models

Ford diesel engines are fitted with Simms fuel injector assemblies. The injector nozzles have four 0.010 or 0.011 diameter holes spaced in a spiral pattern at 90 degrees apart around the tip of the nozzle. Injector nozzles having the 0.010 spray holes are identified by the Simms part No. NL-123 etched on large diameter of nozzle body and nozzles having 0.011 spray holes have Simms part No. NL-141 etched on nozzle body.

All 144 cu. in. diesel engines and 172 cu. in. diesel engines after tractor Serial No. 82582 of 1958-62 production are equipped with injector nozzles having 0.011 spray holes and 172 cu. in. diesel engines prior

Fig. FO102 — Sectional view of the four orifice Simms injector which is used in the Ford diesel engine.

LEAK OFF PIPE CONNECTION
ADJUSTING NUT
CAP NUT
PLATE
COPPER WASHER
SPRING
CAP
FUEL INLET ADAPTER
NOZZLE HOLDER
COPPER WASHER
SPINDLE
NOZZLE RETAINING NUT
NOZZLE AND NEEDLE VALVE ASSEMBLY
NEEDLE VALVE SEAT

to tractor Serial No. 82583 were equipped with nozzles having 0.010 spray holes.

Although there is little difference between the NL-123 and NL-141 nozzles, engine imbalance can result from installing mixed sets of injector nozzles in the same engine. If renewing a complete set of four injector nozzles, install the NL-141 (0.011 spray holes) nozzles. If renewing only one, two or three nozzles, be sure the part number on all nozzles are the same. The NL-123 (0.010 spray holes) nozzles should only be used with Ford part No. 310828 injector pump on early 172 cu. in. diesel engines.

131. TESTING AND LOCATING A FAULTY NOZZLE. If engine does not run properly and a faulty injection nozzle is indicated, such a nozzle can be located as follows: With engine running, loosen the high pressure line fitting on each nozzle holder in turn, thereby allowing fuel to escape at the union rather than enter the cylinder. As in checking spark plugs in a spark ignition engine, the faulty unit is the one which, when its line is loosened, least affects the running of the engine.

132. NOZZLE TESTER. A complete job of testing and adjusting the nozzle requires the use of a special tester, such as shown in Fig. FO101. The nozzle should be tested for opening pressure, spray pattern, seat leakage and leak back.

Operate the tester until oil flows and attach nozzle and holder assembly. Close the tester valve and apply a few quick strokes to the tester handle. If undue pressure is required to operate the lever, the nozzle valve is plugged and should be serviced as in paragraph 138.

133. OPENING PRESSURE. While operating the tester handle, observe the gage pressure at which the spray occurs. This gage pressure should be 2700-2800 psi. If pressure is not as specified, remove the cap nut (Fig. FO102) and turn adjusting nut as required. If opening pressure cannot be brought to within the specified limits (2700-2800 psi), overhaul nozzle as outlined in paragraph 138.

134. SPRAY PATTERN. Operate the tester handle slowly as possible to operate the injector and observe the spray pattern. All four (4) sprays must be similar and spaced approximately 90 degrees to each other in a nearly horizontal plane. Each spray must be well atomized and should spread into a two inch cone at a six inch distance from injector. If spray pattern is not as outlined, overhaul the nozzle as outlined in paragraph 138.

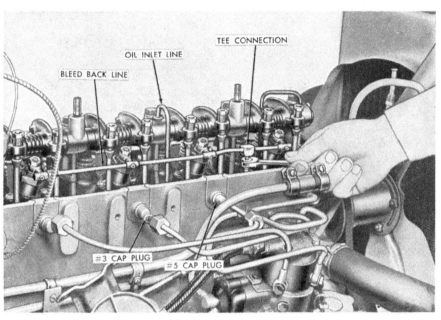

Fig. FO104 — View showing the excess (bleed back) fuel line which must be removed prior to the removal of injectors.

Fig. FO105 — Injectors can be pulled after push rods have been disengaged from rocker arms and moved sideways away from injectors as shown.

Fig. FO106 — Special tool DDN17098 can be used to pull injectors. If injectors are being pulled while fuel tank is on tractor it may be necessary to bend handle of tool to prevent it striking tank.

135. SEAT LEAKAGE. Wipe nozzle tip dry with clean blotting paper then operate tester handle and bring pressure to 150 psi below opening pressure and hold this pressure for one minute. Apply a clean piece of blotting paper to nozzle tip as shown in Fig. FO103. The fuel oil stain should not exceed ½-inch in diameter. If above conditions are not met overhaul injector as outlined in paragragh 138.

136. NOZZLE LEAK BACK. Operate tester handle and bring gage pressure to 2200 psi and note the amount of time it takes for the pressure to drop from 2200 psi to 1500 psi. This time should be between six and forty five seconds.

If elapsed time is not as specified, nozzle should be cleaned and/or overhauled as outlined in paragraph 138.

Note: A leaking tester connector, check valve or pressure gage will show up in this test as excessively fast leak back. If, in testing a number of injectors, all fail to pass this test, the tester rather than the injectors should be suspected.

137. **REMOVE AND REINSTALL.** Remove valve rocker cover and gasket which, due to space limitations, can be facilitated by removing the two cover to cylinder head studs. Tilt top of rocker cover out, push from opposite side of engine with long screwdriver or punch inserted between muffler and inlet manifold to remove.

Disconnect the excess fuel line at the top of each injector then remove line by unscrewing the hex fitting at the "tee" connection in the line to the cylinder head. See Fig. FO104. Valve push rods interfere with the removal of the injectors and can be repositioned as follows: Place a ½-inch box wrench over head of adjusting screw and lift the rocker arm enough to release the valve push rod which can then be moved sideways away from injector as shown in Fig. FO105.

NOTE: Fig. FO105 shows all four injectors removed at once. Unless sending the injectors to a service station, it is usually more practical to remove only one injector at a time. Turn the engine so that the push rods at each side of the injector are free, then lift the rocker arms using a box end wrench on adjusting screw as lever; if rocker arm cannot be lifted high enough to move push rod aside, valve is contacting piston and engine must be turned slightly. Care should be taken in turning the engine with push rods disconnected from rocker arms as it is possible to bend the push rods.

Carefully clean all dirt and foreign material from the pressure line connections at the injectors and disconnect the lines. Cap off lines and the injector inlets. Injectors can now be pulled by using special tool No. DDN 17098 as shown in Fig. FO106. CAUTION: Be sure that legs of tool bear on flange of cylinder head and not on the rubber seal. Remove injector seat washer by using a bent wire. If washers are stuck in bore, the following method of removal can be used. Grind the blade of a screwdriver to a taper that will permit it to protrude not more than ⅛-inch through the I.D. of seat washer. Place blade of screwdriver through I.D. of stuck washer and rap the handle sharply so blade will bite into washer. Washer can now be twisted free.

Always use new injector seat washers when reinstalling injectors and make sure that injector bores in cylinder head are absolutely clean and that washers are flat in bore. Use a light and mirror if necessary. Install each injector by positioning the nozzle end in its bore in cylinder head and guiding the injector rubber seal into its slot. Work each injector downward, using hands only. DO NOT use a hammer. Align injector and install mounting bolts. Torque each bolt to 15 Ft.-Lbs. Reposition the valve push rods to their proper place. Reinstall the excess fuel line by reversing the removal procedure. Note: Some early models may have flat brass washers under heads of capscrews retaining excess fuel line to injectors. Others use special capscrews with sealing ridge on bottom of head. Flush pressure lines and injectors prior to connecting the lines by cranking engine with starter and directing the fuel discharge from pressure lines into injector inlets. After all lines and injectors have been flushed, connect lines and tighten, then back off two turns. Set and throttle to idle position and crank engine with starter. When fuel spits out at connections, tighten same. Start engine and bring to operating temperature, then set valve clearance to 0.014-0.016. Reinstall valve cover and gasket and use a non-hardening sealent at corners of grommets where they contact the cylinder head and rocker cover gasket.

138. **OVERHAUL.** Unless complete and proper equipment is available (Bacharach 66-5040A or equivalent), do not attempt to overhaul diesel nozzles.

Refer to Fig. FO102 and proceed as follows: Secure injector holding fixture in a vise and mount injector in fixture. **Never** clamp the injector body in a vise. Remove the cap nut and back off adjusting nut, then lift off the upper spring disc, injector spring and spindle. Remove the nozzle retaining nut using Bacharach special tool 66-1078, or equivalent, and remove the nozzle and valve. Nozzles and valves are a lapped fit and must never be interchanged. Place all parts in clean fuel oil or calibrating fluid as they are disasembled. Clean injector assembly exterior as follows: Soften hard carbon deposits formed in the spray holes and needle tip by soaking in a suitable carbon solvent, then use a soft wire brush to remove carbon from the needle and nozzle exterior. Rinse the nozzle and needle immediately after cleaning to prevent the carbon solvent from corroding the highly finished surfaces. Clean the pressure chamber of the nozzle with a 0.040 reamer as shown in Fig. FO-107. Clean the spray holes in the nozzle with the proper sized wire probe held in a pin vise as shown in Fig. FO108. To prevent breakage, the wire probe should protrude from pin vise only far enough to pass through spray holes. Rotate pin vise without applying undue pressure. Use a 0.009 wire for 0.010 injector spray holes and a 0.010 wire for 0.011 spray holes.

The valve seats are cleaned by inserting the small end of a valve seat scraper into the nozzle and rotating. Then reverse the scraper and clean the upper chamfer with the large end. See Fig. FO109. The annular groove in top of the nozzle and the pressure chamber are cleaned by using (rotating) the pressure chamber tool as shown in Fig. FO110.

With the above cleaning accomplished, back flush nozzle and needle by installing the reverse flushing adapter on the injector tester and positioning the nozzle and valve in adapter tip end first. Secure with the knurled nut and rotate the needle in the nozzle while flushing to make sure it is free. After nozzle is back flushed, the seat can be polished by using a small amount of tallow on the end of a polishing stick and rotating the nozzle as shown in Fig. FO111.

If the leak-back test time was greater than 45 seconds (paragraph 136), or if the valve is sticking slightly, correction can be made by remating the needle and nozzle assembly. This is accomplished by using a polishing compound (Bacharach No.

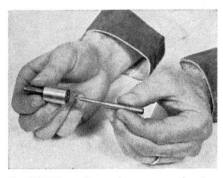

Fig. FO107 — Clean the pressure chamber of the nozzle with a 0.040 reamer as shown.

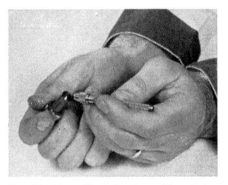

Fig. FO108 — View showing orifices being cleaned with a wire probe held in a pin vise. Refer to text.

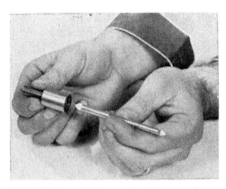

Fig. FO109 — Valve seat and upper chamfer are cleaned by using scraper as shown. Refer to text.

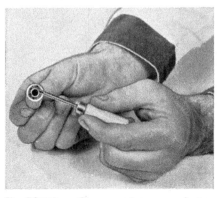

Fig. FO110 — Clean annular groove in top of nozzle by using pressure chamber tool as shown.

66-0655) consisting of tallow and a small amount of very fine lapping compound and proceeding as follows: Hold needle in a chuck and polish same using a piece of felt coated with a very small amount of the above mentioned special compound; or, place the nozzle in the chuck of a drill having a maximum rpm of not more

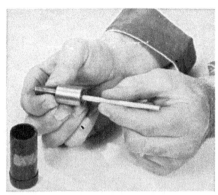

Fig. FO111 — Nozzle seat can be polished by using a small amount of tallow on a polishing stick and rotating nozzle as shown.

Fig. FO112 — Needle and nozzle assembly can be remated by using an electric drill and polishing compound as shown. Refer to text.

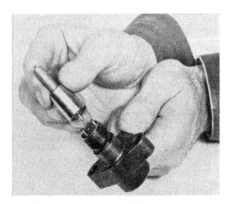

Fig. FO113 — When positioning nozzle and needle on body be sure that body dowel pins are correctly aligned in nozzle.

than 450, then apply a small amount of special compound on the needle valve and insert same in nozzle body. Turn nozzle to lap and be sure to hold needle up off the pressure chamber shoulder during operation to avoid damage to needle. See Fig. FO112. Care should be taken to see that lapping compound does not damage the needle seat. Back flush and clean assembly.

Before reassembling, rinse all parts in clean fuel oil or calibrating fluid and install while still wet. The injector inlet adapter normally does not need to be removed. However, if adapter is removed, use a new copper washer when reinstalling. Position the nozzle and needle valve on injector body and make sure dowel pins in body are correctly located in nozzle as shown in Fig. FO113. Install the nozzle retaining nut and torque to 50 Ft.-Lbs. Note: Place injectors in holding fixture to torque nut. Install the spindle, spring, upper spring disc and spring adjusting nut. Tighten the adjusting nut until pressure from spring is felt. Connect the injector to the nozzle tester and adjust opening pressure to 2700-2800 psi. Use a new copper gasket and install cap nut. Recheck nozzle opening pressure to see that it has not changed.

Retest the injector as outlined in paragraphs 133 through 136, and if injector fails to pass these tests, renew the nozzle and needle.

Note: If injectors are to be stored, it is recommended that they be cleaned in calibrating fluid prior to storage.

INJECTION PUMP

All Diesel Models

The fuel injection pump is a self-contained unit which includes the engine governor and components for delivering fuel in the properly metered amounts. The pump is vertically mounted on the right side of the engine block and is driven by the engine camshaft. Other than renewing the assembly as a complete unit, the only servicing requirements are timing, speed adjustment and the renewal of the pump drive shaft and seals and pump drive gear.

139. **PUMP TIMING.** Before checking or adjusting injection pump timing, refer to the nameplate attached to the pump housing for the Ford part number stamped thereon. Recommended timing specifications (in degrees of flywheel rotation BTDC on No. 1 cylinder) for the different injection pumps that have been used are as follows:

Fig. FO114 — Procedure for timing early 172 cubic inch diesel engine. Method of timing other models is identical except for flywheel timing marks. Pump timing marks can be aligned by loosening nuts (1) and rotating pump.

Fig. FO115 — Remove injection pump drive gear by turning clockwise with an end wrench or screwdriver.

144 Cu. In. Engine:

Ford Part No.	Timing
B9NN-9A543-A	26° BTDC
CONN-9A543-C	23° BTDC
CONN-9A543-H	23° BTDC

172 Cu. In. Engine:

Ford Part No.	Timing
310828	18° BTDC
B9NN-9A543-B	18° BTDC
CONN-9A543-D	23° BTDC
CONN-9A543-J	23° BTDC

NOTE: If the pumping elements in earlier production pumps (Ford part Nos. B9NN-9A543-A for 144 cu. in. engine and 310828 and B9NN-9A543-B for 172 cu. in. engine) have been renewed using the latest type service parts, injection timing should be adjusted to 23 degrees BTDC instead of the specification given for the particular injection pump. If it is not known whether the pump is equipped with the latest type pumping element parts or not, timing should be set to either 23 degrees BTDC or at timing specified for that pump part number, whichever provides best engine operation.

CAUTION: Some mechanics prefer to adjust the injection pump timing by placing tractor on a PTO dynamometer and adjusting timing for optimum power output. If this method is used, do not attempt to change the injection timing while engine is running. Stop the engine, advance or retard timing, and then recheck power output. Loosening the injection pump mounting bolts while engine is running may cause pump seizure and subsequent serious damage.

To time the fuel injection pump, turn engine in normal direction of rotation until No. 1 piston is coming up on compression stroke; then, continue turning engine slowly until the correct flywheel timing mark is aligned with index mark on timing port (See Fig. FO114). Shut off fuel

supply valve, remove the timing window cover from side of fuel injection pump and check to be sure that the scribe lines on cam and governor cage are exactly aligned. If not, loosen the two pump mounting stud nuts (1) and shift pump housing until scribe marks are aligned; then, tighten the pump mounting stud nuts.

140. INJECTION PUMP R&R. To remove the complete injection pump unit, first shut off fuel supply and thoroughly clean dirt from pump, fuel lines and connections. Remove timing window from side of fuel injection pump. Turn crankshaft in normal direction of rotation (clockwise) until correct degree BTDC timing mark on flywheel is in register with index mark at timing port and the scribe lines on cam and governor cage of fuel injection pump are aligned. Disconnect fuel supply line and throttle linkage from pump. Remove pressure lines from injectors and mounting nuts, then lift pump and drive shaft as an assembly from engine. If necessary, pressure lines can now be removed from injection pump.

Before reinstalling the pump, remove the timing hole cover from side of injection pump and turn the pump drive shaft until timing lines are aligned as shown in Fig. FO114. Mount pump to engine. Connect the engine controls and fuel supply line. Before connecting the pressure lines flush the same with clean fuel oil. Recheck the pump timing as in paragraph 139 and bleed the fuel system as in paragraph 130.

141. INJECTION PUMP DRIVE SHAFT. The injection pump drive shaft is fitted with "U" cup seals which may be renewed. In addition, the shaft is available for service.

With injection pump removed as in paragraph 140, pull pump drive shaft from housing. The shaft is retained in the housing by an "O" ring located on the shaft above the seals. Inspect seals for cuts or tears and renew if necessary.

Notice the offset dimple on the end of the tang which fits into the distributor rotor of pump. A similar dimple will be found in the slot of the distributor rotor. When reinstalling the pump shaft align these two marks so that pump can be properly timed to engine.

Prior to installing pump shaft, coat the "U" cup seals and "O" ring with Lubriplate or its equivalent. Position shaft so the previously mentioned dimple marks will mate and insert shaft into pump housing. Work the seals into the bore using a pencil or blunt tool. DO NOT force the seals. When shaft is properly positioned the drive tang shoulder will be approximately $\frac{1}{16}$-inch below the top of the shaft bore in pump housing.

Reinstall the injection pump as in paragraph 140, then time as in paragraph 139. Bleed fuel system as in paragraph 130.

142. PUMP DRIVE GEAR. The pump drive gear is driven by the engine camshaft. Normally, this gear is removed only when it needs renewing or when engine work requires its removal.

Gear is removed by placing an open end wrench or a wide blade screw driver in the slot and turning gear clockwise. This raises and unmeshes gear so it can be lifted from engine block. See Fig. FO115.

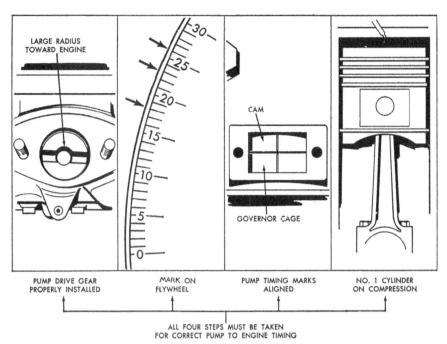

Fig. FO116 — When timing the injection pump to engine all four steps shown must be done. Note position of large offset of pump drive gear.

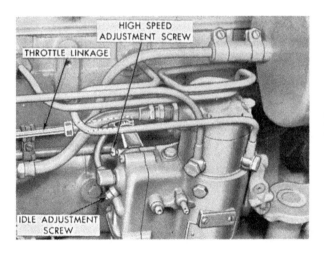

Fig. FO117 — View showing location of the speed adjusting screws and throttle linkage. Refer to text to set throttle dead travel.

While gear can be installed in any position, only one position will insure correct pump to engine timing. Crank engine in normal direction of rotation until number one piston piston is coming up on compression stroke, then continue turning crankshaft until the appropriate timing mark on flywheel registers with index mark at timing port. Refer to paragraph 139. Place drive gear in the bore of the engine block so that the slot in the top of gear is in the vertical position and the large radius of offset is on the right hand side. Rotate the gear ¼-turn counter-clockwise. When properly meshed, the slot in the top of the drive gear will be in the horizontal position and the large radius of the offset will be toward engine as shown

in Fig. FO116. It may be necessary to put the tractor in gear and rock it backward and forward for complete meshing of pump drive gear with camshaft and oil pump drive shaft.

143. GOVERNOR-ADJUST. To adjust governor, start engine and bring to operating temperature. Position hand throttle lever for low idle which should be 675 rpm. If low idle is not as specified, refer to Fig. FO117, and proceed as follows: Loosen jam nut on the idle speed adjusting screw and using an Allen wrench, turn adjusting screw IN to increase or OUT to decrease engine rpm. Tighten jam nut and recheck.

With low idle speed properly adjusted, move hand throttle lever to

wide open position and check engine rpm which should be 2450 for tractors equipped with five-speed or "Select-O-Speed" transmissions, or 2250 for tractors equipped with four-speed transmissions. If engine rpm is not as specified, loosen the jam nut on the high speed adjusting screw (See Fig. FO117) and using a screwdriver, turn screw IN to decrease or OUT to increase engine rpm. Tighten jam nut and recheck.

144. THROTTLE LINKAGE. The throttle linkage must be adjusted to provide ¼-½-inch dead travel of the hand throttle lever at the idle position whenever a new injection pump or new linkage is installed. Throttle linkage dead travel should also be checked when making normal service adjustments for engine speed.

Adjust throttle linkage as follows: With engine running, set the hand throttle lever to the idle position and disconnect throttle linkage at the injection pump operating lever. See Fig. FO117. Pull the pump operating lever rearward to its idle position. NOTE: There is no stop at this position, however; when this position is reached a slight drag will be felt on the lever. Pulling the lever further back and over-riding this drag, will result in the lever contacting the stop and shutting off the fuel.

With the pump operating lever in its idle position, adjust the throttle linkage by loosening the jam nut of the ball cage and turning cage until it is aligned with the ball on the pump operating lever. Connect the linkage and tighten jam nut.

With engine idling at 675 rpm, move the hand throttle lever about ¼ to ½-inch off its idle stop position, at which time the engine should just start to accelerate. Repeat linkage adjustments until the hand throttle lever dead travel is established. This dead travel is entirely controlled by the length of the throttle linkage and must be obtained to prevent activating the injection pump governor in low idle range.

ACCELERATOR PEDAL—ADJUST

Industrial Models

145. Adjust the governor and hand throttle linkage as outlined in paragraphs 143 and 144. Set the hand throttle in the idle position, fully depress the accelerator pedal, then re-

GOVERNOR
(Non-Diesel)

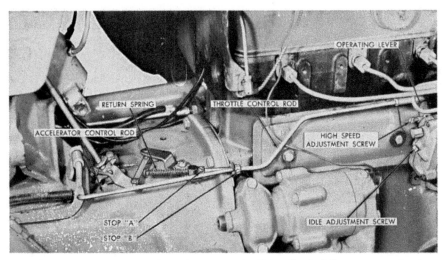

Fig. FO118 — Accelerator pedal linkage on series 1801 diesel. Adjust return spring tension at stop "A" and high idle speed at stop "B". Other industrial models are similar.

Non-diesel tractor models use a variable speed, centrifugal governor which is mounted on front of crankshaft. In early model tractors, the governor fork and lever was mounted in a separate governor housing attached to timing gear cover. In 1955, the separate cover housing was eliminated, and governor fork and lever is contained in timing gear cover. The separate governor housing is no longer available for service; if renewal is indicated, install the new type timing gear cover.

Prior to early 1958, a flyball type (Novi) governor unit was used. The governor unit was then changed to a flyweight (Pierce) type. Except for the disassembly of the weight unit, overhaul procedures are similar.

NOTE: The flyweight type (Pierce) governor unit is not splined to the crankshaft, but is driven by the friction of the crankshaft pulley. If governor action is sluggish or erratic, check to see that pulley retaining cap screw is tight. Tighten the cap screw to a torque of 100 Ft.-Lbs. and recheck performance before overhauling governor. On flyball governor, a dimple in ID of driver fits in crankshaft spline.

lease it to see if engine speed quickly returns to the idle position.

If the engine speed does not quickly return to the idle position, examine the acelerator pedal linkage and throttle control rod (See Fig. FO118) for binding. Eliminate any binding, by realignment and lubrication. If sluggish action is still noted, increase tension on return spring by moving "Stop 'A'."

Note: If it is found necessary to reposition Stop A, linkage will need to be readjusted as outlined in paragraph 144.

To adjust the accelerator pedal high speed setting, stop the engine after proper hand throttle adjustment, move the hand throttle to full throttle speed and loosen Stop "B". Depress the accelerator pedal against the step plate and move Stop "B" on the throttle control rod until it firmly contacts the accelerator control rod. This should allow full engine speed by depressing the accelerator pedal, regardless of the hand throttle setting.

146. SPEED ADJUSTMENT. Before attempting to adjust the governed speed, first free-up and align all linkage to remove any binding tendency. Disconnect governor to carburetor link rod (R—Fig. FO120) at the carburetor throttle lever. While holding the carburetor throttle lever in the wide open position, move the hand throttle to the wide open position and note whether link connector will slip over the ball on the carburetor throttle lever. If the link connector will not slip over the ball, adjust the clevis (L) until proper condition is obtained.

Reconnect the link rod (R), start engine and bring to operating temperature. Adjust the throttle stop screw on carburetor to obtain an engine slow idle speed of 450-475 rpm. At idle speed setting, tip of throttle rod end assembly (C) should be against governor arm with throttle lever fully forward. If there is a gap between governor arm and throttle rod end assembly, loosen jam nut (N) retaining throttle rod end and screw end assembly out until contact is made with governor arm. Retighten jam nut. Move the hand throttle lever to the wide open position and check

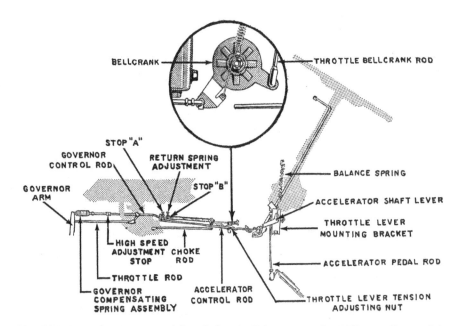

Fig. FO119 — Accelerator pedal and throttle linkage on series 1801 gasoline models. Other industrial models are similar.

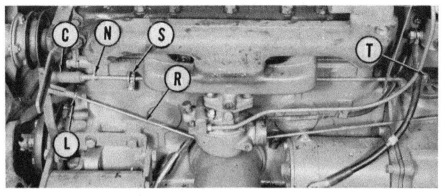

Fig. FO120 — View of left side of non-diesel engine showing governor rod compensating end assembly (C), adjusting lock nut (N), stop (S) and linkage friction adjusting nut (T). Carburetor to governor lever rod is (R) and rod length adjusting lock nut is (L).

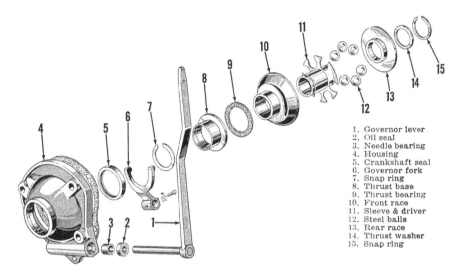

1. Governor lever
2. Oil seal
3. Needle bearing
4. Housing
5. Crankshaft seal
6. Governor fork
7. Snap ring
8. Thrust base
9. Thrust bearing
10. Front race
11. Sleeve & driver
12. Steel balls
13. Rear race
14. Thrust washer
15. Snap ring

Fig. FO121 — Exploded view of early production Novi flyball governor unit. Separate governor housing (4) is no longer available; renew with one-piece governor housing and timing gear cover (1—Fig. FO122). Needle bearing (3) is not serviced.

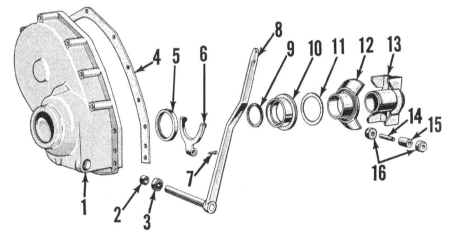

Fig. FO122 — Exploded view of late production Pierce flyweight governor unit. May be installed in early production tractors in place of Novi flyball unit shown in Fig. FO121.

1. Timing gear cover	5. Crankshaft seal	9. Snap ring	13. Sleeve & driver
2. Bushing or needle bearing (not serviced)	6. Governor fork	10. Front race	14. Pin
3. Oil seal	7. Retaining screw	11. Thrust bearing	15. Governor weight
4. Gasket	8. Governor lever	12. Rear race	16. Weight rollers

the high idle, no-load speed which should be 2200-2250 rpm on tractors with four-speed transmissions and 2400-2450 rpm on tractors with five-speed or "Select-O-Speed" transmissions. If the high idle, no-load speed is not as specified, loosen stop (S) and move the hand throttle until the tractor "Proof-Meter" registers the specified rpm. Then, secure the stop (S) against the throttle rod guide bracket by tightening the lock nut.

If the hand throttle tends to creep toward closed position, correct the trouble by tightening the castellated nut (T) which will put more tension on the bell crank spider spring washer.

147. **ACCELERATOR PEDAL ADJUSTMENT.** To adjust the accelerator pedal on the industrial tractors, first adjust the governor and throttle linkage as outlined in paragraph 146, set the hand throttle in the idle position and fully depress the accelerator pedal. The engine speed should be the same as for the full throttle setting with the hand throttle. If it is not, adjust the length of the accelerator pedal rod (Fig. FO119) until the accelerator pedal contacts the step plate as the engine speed reaches the maximum setting.

Move the foot quickly off the accelerator pedal and note if the engine immediately responds by returning to the idle position. If it does not, check the accelerator pedal linkage and remove any binding or tightness. If engine is still sluggish in returning to idle speed, increase tension on return spring by moving Stop "B" to the rear.

Acelerator pedal pressure can be changed by repositioning the balance spring in one of the three holes in the accelerator shaft lever.

148. **REMOVE AND REINSTALL.** The governor weight unit can be removed from the engine crankshaft after removing the timing gear cover as outlined in paragraph 96.

149. **OVERHAUL.** To remove the governor shaft and lever assembly, remove locking screw or roll pin from fork and withdraw shaft and lever from fork and governor housing or timing gear cover. Pry out governor shaft oil seal and install new seal with lip to inside using drift of correct size. Governor shaft needle bearings or bushing is not available for service. The crankshaft front oil seal can be renewed without disassembling

governor housing or timing gear cover. Install new seal with lip facing inward and make sure seal is fully bottomed.

To disassemble the removed early type flyball weight unit, remove the two snap rings (7 and 15—Fig. FO-121). Check the balls, sleeve and races for wear, pitting and scoring. Parts must work freely without binding.

To disassemble the removed weight unit of the late type flyweight governor, remove snap ring (9—Fig. FO-122) and separate parts. Check thrust bearing (11) and the bearing contacting surfaces of the race (10) and the fork base (12).

NOTE: The early type governor assemblies can be replaced with the late type governor assemblies by using the following new components: Centrifugal assembly, lever assembly, governor fork and governor compensating rod end assembly.

Fig. FO123 — To obtain access to radiator, to front support or to hydraulic power package reservoir, grille housing must be lifted off as shown.

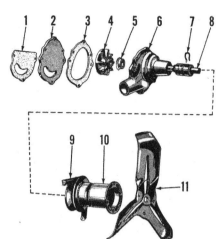

Fig. FO124 — Exploded view of water pump assembly.

1. Gasket	7. Retaining wire
2. Rear plate	8. Shaft & bearing assy.
3. Gasket	9. Fan belt
4. Impeller	10. Water pump pulley
5. Shaft seal	11. Fan blade assy.
6. Pump body	

COOLING SYSTEM

The radiator is fitted with a 3.5-4.5 pound pressure cap which raises the cooling system boiling point. The thermostat on gasoline models starts to open at 157-162 degrees F. and is fully open at 177-182 degrees F. On diesel models the thermostat starts to open at 177-182 degrees F. and is fully open at 197-202 degrees F.

RADIATOR

All Models Except Heavy Duty Industrial

150. To remove the radiator, first drain cooling system and remove hood. See paragraph 81. Remove the radiator hoses, unbolt radiator from the front support and lift radiator from tractor. On Row Crop models, remove fan blades to provide clearance when radiator is lifted off.

Heavy Duty Industrial

151. **R&R GRILLE.** To remove the grille, first drain the hydraulic power package oil system at the system filter on right side of tractor engine. Remove the grille front door and medallion, disconnect the headlight wires and disconnect the two hoses connecting the pump inlet and system return

line to the hydraulic reservoir. Remove the six Phillips head cap screws securing the hood side panels to the grille and remove the nuts from the six bolts securing the grille to the tractor front support. Attach a rope or chain to the grille as shown in Fig. FO123 and remove grille with a suitable hoist.

152. **R&R RADIATOR.** Drain the cooling system and power steering system and remove the tractor grille as outlined in paragraph 151. Disconnect the upper and lower radiator hoses and remove the two nuts securing the radiator to the tractor front support. Pull top of radiator forward to gain access and disconnect the two oil cooler lines at the bottom rear of the radiator; then, lift the radiator from tractor.

Reinstall radiator by reversing the removal procedure.

WATER PUMP

All Models

153. To remove the water pump, drain cooling system and loosen the fan belt. Disconnect the radiator hose from pump and remove the fan blades unit. On some models, power steering pump must be removed to provide

clearance. Remove the cap screws retaining the pump to the cylinder block and withdraw the pump from engine.

To disassemble the pump, refer to Fig. FO124 and proceed as follows: Using a suitable puller, remove the pulley (10). Remove the pump cover (2) and bearing retaining ring (7). Support the pump body in a suitable manner and press the shaft and bearing assembly out of the impeller and pump body. Remove the seal assembly by driving it toward impeller end of body. Examine sealing surface of impeller. If the surface is scratched or pitted, renew the impeller.

When reassembling, press the shaft and bearing assembly into the housing until the ring groove in bearing is in register with the groove in the housing and insert retainer in groove. Press the seal assembly into the housing, using an adaptor that will press against the shoulder of the seal. Install the impeller so that rear surface of impeller is 0.015-0.025 less than flush with rear surface of pump housing. This clearance is necessary to prevent impeller from rubbing the rear cover plate. Install the pulley in a press while supporting rear end of shaft (8). Make certain that pulley hub does not bind on the housing. Install the rear cover plate.

When installing the pump on the engine, tighten the retaining cap screws to a torque of 25-35 Ft.-Lbs.

IGNITION AND ELECTRICAL SYSTEM

GENERATOR AND REGULATOR

Non-Diesel Models

154. A Ford designed, shunt-wound, two-pole type generator is used. Generator output is controlled by a two-unit type regulator consisting of a cut-out relay and a voltage regulator. Ford part numbers assigned to these units are FAC-10002-A for the generator, FAG-10505-A for the regulator. Specifications are as follows:

Generator:
Armature part No. ...FAA-10005-B
Engine rpm test voltage......1200
Watt 160
Renew brushes if
  shorter than ....½-inch
Brush spring tension..26-34 ounces
Max. generator
  output ......20 amps @ 1650 rpm
Field part number....81A-10175-A
Field resistance, ohms..3.2 @ 70° F
                        3.8 @ 140° F

Regulator:
Cutout voltage .....opening, 5.1-5.5
                    closing, 6.0-6.6
Voltage limiter voltage 7.1-7.5 @ 75° F
Reverse current at min.
  of 6 volts............6 amps. max.

Diesel Models

155. Ford designed, shunt-wound, two-pole type generator is used. Generator output is controlled by a two-unit type regulator consisting of a cutout relay and a voltage regulator. Ford part numbers assigned to these units are 2900450 and B9NF-10002-A for the generator; 2900446 and CONF-10505-A for the regulator. Specifications are as follows:

2900450 Generator:
Armature part No. ...FAA-10005-B
Engine rpm test voltage......1500
Watt 300
Renew brushes if
  shorter than ....½-inch
Max. generator
  output ......20 amps @ 1500 rpm
Brush spring tension..32-40 ounces
Field part No. FAS-10175-B
Field resistance, ohms...7.2 @ 70° F

2900446 Regulator:
Cutout
  voltage Opening 11.0-13.0 @ 75° F
           closing 12.2-13.0 @ 75° F
Voltage limiter
  voltage ....14.6-15.4 @ 75° F
Reverse current at min.
  of 12 volts.........6.0-9.5 amps

B9NF-10002-A Generator:
Armature part No. ...FGT-10005-A
Engine rpm test voltage......1500
Watt 300
Renew brushes if shorter
  than ....½-inch
Max. generator
  output ......25 amps @ 1500 rpm
Brush spring tension..32-40 ounces
Field part No. 2900619
Field resistance, ohms..7.2 @ 70° F

CONF-10505-A Regulator:
Cutout
  voltage opening 11.0-13.0 @ 75° F
           closing 12.2-13.0 @ 75° F
Voltage limiter
  voltage ....14.6-15.4 @ 75° F
Reverse current at min.
  of 12 volts.........6.0-9.5 amps

STARTING MOTOR

Non-Diesel Models

156. The 6-volt Ford designed starting motor is provided with a "follow-through" type drive. The unit should draw 100 amps at no-load and 100-150 amps when cranking a warm engine at 100 rpm. Brush spring tension should be 48-56 ounces.

Diesel Models

157. The 12-volt Ford designed starting motor is provided with the same "follow-through" type as the 6-volt non-disel units. The unit should draw 100 amps at no load. Ampere load when cranking a warm engine is 190-225 at a cranking speed of 180-230 rpm. Brush spring tension should be 50-58 ounces.

IGNITION SYSTEM

Non-Diesel Models

158. Auto-Lite ATL3A spark plugs are used on LP-Gas models; Auto-Lite AL7T are used on all other models. Plug electrode gap is 0.028-0.033 for LP-Gas models, 0.025-0.029 for all other models. Plug tightening torque is 24-30 Ft.-Lbs. The 7RA-12300-C condenser has a capacity of 0.21-0.25 mfd. The 8BA-12029 coil has a primary resistance of 1.05-1.15 ohms at 75 degrees F., and a secondary resistance of 4100 ohms at 75 degrees F. The FDN-12127-A distributor is used on all models prior to 1958. Models

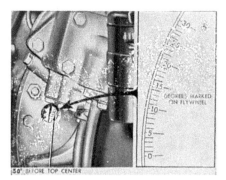

Fig. FO125 — View showing flywheel marking, each line representing one degree.

after 1957 use distributor number 311185. Overhaul of both units is conventional.

159. **IGNITION TIMING.** Breaker contact gap is 0.024-0.026. Firing order is 1-2-4-3. To install and time the distributor, first adjust the breaker contact gap to 0.025 and proceed as follows: Remove the number one spark plug and the timing hole cover from right side of engine. Refer to Fig. FO125. Crank engine until the number one piston is coming up on compression stroke; then, using a screw driver as a pry bar on the flywheel teeth, turn flywheel in normal direction until the specified fully advanced timing mark on flywheel is in register with the index mark at timing port. The fully advanced timing marks are as follows:

1955-57 models ....26°
1958 and later models..........24°

With crankshaft in this position, turn the distributor shaft until the rotor arm is in number one firing position, mount distributor on engine and tighten the clamp screw finger tight. Rotate the breaker cam in the normal direction as far as it will go and while holding the cam in this position, turn the distributor body in the opposite direction until the breaker contacts just start to open; then tighten the distributor clamp securely.

When checked with a timing light and engine running at 2200 rpm, timing marks mentioned above should register with the index at the timing port. With the engine running at the slow idle speed of 450-475 rpm, the 4° timing mark should register with the index at timing port.

If the fully advanced timing is O.K., but the idle timing is not, a malfunctioning distributor advance governor (cam and weights assembly) is indicated.

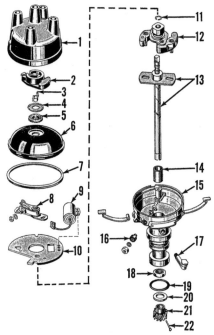

Fig. FO126 — Exploded view of ignition distributor assembly. Service shaft (13) is not drilled for gear retaining rivet (22).

1. Distributor cap	12. Adance weights
2. Rotor	13. Distributor shaft
3. Rotor spring	14. Bushing
4. Seal retainer	15. Distributor housing
5. Felt seal	16. Insulator
6. Dust cap	17. Oiler
7. Gasket	18. Oil seal
8. Breaker points	19. "O" ring
9. Condensor	20. Thrust washer
10. Breaker plate	21. Drive gear
11. Snap ring	22. Rivet

160. TEST AND OVERHAUL. Breaker contact gap is 0.024-0.026. Cam dwell angle is 58-63 degrees. Breaker arm spring tension is 17-20 oz. To increase the tension, loosen the condenser lead nut and slide the leaf spring in direction toward point end of breaker arm so as to decrease the radius of the spring at the arm pivot.

The distributors are similar on all models except that a different cam and weight assembly is used on 1958 and later models than on 1955-57 models. Assuming the distributor is off the engine and being tested on a synchroscope, the advance data for the two different cam and weight assemblies in distributor degrees and distributor rpm are as follows:

FAA-12176-A (1955-57)

	Degrees @ R.P.M.
Start advance	0 @ 225-235
Intermediate adv. .	4.75- 5.75 @ 600
Full advance	10.5-11.5 @ 1000

311184 (1958 and Later)

Start advance	0 @ 225-235
Intermediate advance	4.5 @ 600
Full advance	10 @ 1100

If the distributor is being checked on the tractor using a timing light, and the ignition timing is correctly adjusted, the advance data for the two different breaker cam and weight assemblies, in engine rpm and flywheel degrees, are as follows:

FAA-12176-A (1955-57 Models)

450-475 RPM	4°
1200 RPM	13½-15½°
2000-2200 RPM	25-27°

311184 (1958 and Later Models)

450-475 RPM	4°
1200 RPM	13°
2200 RPM	24°

161. The following points should be observed when overhauling the distributor unit: To remove the shaft from the distributor, file off and remove the drive gear retaining rivet (22—Fig. FO126) and drift off the gear (21). Always renew the shaft oil seal and "O" ring. The drive shaft bushing (14) should be renewed if running clearance exceeds 0.003 and should be final sized after installation. Upper end of bushing should be flush with top of spot face in distributor body. Before riveting the gear to drive shaft, check end play of shaft which should be within the limits of 0.026-0.037. If new distributor shaft is being installed, press gear on shaft until proper end play is obtained, then, using rivet hole in gear as template, drill hole through shaft for rivet. Rivet holes are not drilled in service shafts.

CLUTCH

Tractors without live power take-off may be equipped with either a 9, or a 10 inch single plate, spring loaded, dry disc clutch. Tractors with live power take-off are equipped with a 9 inch dual plate spring loaded clutch. Tractors equipped with the "Select-O-Speed" transmission have no master clutch, the flywheel being fitted with a disc type torque limiting clutch.

Refer to paragraph 166 for dual clutches, and paragraph 230 for the "Select-O-Speed" torque limiting clutch.

TRACTOR SPLIT

Non-Diesel Models Except Heavy Duty Industrial

161. Drain the hydraulic system, remove the air cleaner bowl and unhook governor compensating spring. Disconnect governor compensating rod at bell crank and remove cap screws attaching the throttle control arm bracket to transmission housing. Loosen castellated nut from bell crank stud and remove bell crank and friction disc asembly. Remove choke control rod and starting motor; then remove the remaining cap screws retaining steering gear housing to top of transmission. Disconnect wire from starter switch, tail light wire from transmission and lay wires over the steering gear housing. Remove exhaust pipe and unbolt hydraulic pump manifold from transmission.

On models so equipped disconnect the radius rods at transmision. Note: On those models equipped with power steering it is not necessary to disconnect the power steering cylinders or the lines.

On row crop models, disconnect drag link at rear end.

On "Select-O-Speed" models, remove the drive selector and pto control cable as outlined in paragraph 234 and disengage the traction coupling as outlined in paragraph 223.

Support both halves of tractor separately and remove the bolts and cap screws retaining engine and side rails (if row crop type) to transmission. Carefully raise the engine and at the same time pry up on the steering gear housing until it is free of the locating dowels.

Note: If desired, the steering gear housing can be raised and held in position by inserting wooden blocks between fuel tank and the bottom of the hood sides. Be sure steering gear housing clears dowels when assembly is lowered.

Slide engine forward until front side of steering gear housing meets with the bell crank stud. Continue to raise engine half of tractor until steering gear housing clears the bell crank stud and carefully move the engine forward until the clutch is clear of the clutch housing.

Diesel Models Except Heavy Duty Industrial

162. Drain hydraulic system and remove air cleaner tray. Disconnect battery cables, loosen battery hold down and remove the battery hold down and battery with the cables attached. Work through the hole in the battery tray and disconnect injection pump and throttle lever linkage from

bell-crank. Remove air intake tube, then remove starter solenoid from its mounting but do not disconnect wiring. Now remove battery tray and bell-crank from tractor. Disconnect throttle lever bracket from top of transmission housing. Disconnect Proof-Meter cable at hydraulic pump. Disconnect battery cable from starter and remove starter. Disconnect wires to starter button and tail light.

On models so equipped disconnect radius rods at transmission. Note: On models equipped with power steering it is not necessary to disconnect the power steering cylinder or the lines.

On row crop models, disconnect drag link at rear end.

On "Select-O-Speed" models, remove the drive selector and pto control cable as outlined in paragraph 234 and disengage the traction coupling as outlined in paragraph 223.

Temporarily install the two rear tank support bolts to the tank support and cylinder head. (These bolts were taken out when battery tray was removed). Remove the cap screws retaining the steering gear housing to the transmission, then using a hoist attached to the steering arms, lift steering gear housing until it clears the locating dowels. Hold steering gear in this position by inserting wooden blocks between fuel tank and hood sides at rear. Blocks approximately 1½ inches square and 12 inches long should be sufficient. Lower hoist and check to see that steering gear housing will still clear locating dowels. Use larger blocks if necessary. Position a rope sling around the rear of engine block and attach hoist. Place rags or some other means of protection between ropes and tractor hoods and lift hoist until it just takes the weight. Place a rolling floor jack under transmission from rear. Remove engine to transmission bolts and separate tractor.

Heavy Duty Industrial Models

163. Drain the hydraulic system and, on full power steering models, disconnect steering cable housing from actuator connection and remove cable from actuator assembly by turning steering wheel counter-clockwise until cable is free of actuator wheel. On diesel models, disconnect the battery cables and remove battery. On gasoline models, disconnect throttle linkage and remove the choke rod. On all models, disconnect the accelerator pedal linkage, remove the three bolts

securing the starter, and fasten the starter to the engine adapter plate with a ⅜ NF x ¾ cap screw in the upper hole next to the engine block. On "Select-O-Speed" models, remove the drive selector and pto cable as outlined in paragraph 234 and disengage traction coupling as outlined in paragraph 223.

Support the transmission housing by means of a rolling floor jack, support the two side rails at rear on suitable stands and support the engine by means of a jack under the oil pan.

Remove the four bolts securing the rear fenders and side rails to the rear axle housings, disconnect the light wire and remove the fenders. Unbolt and remove the step plates and the two braces attaching the transmission housing to the side frames. Remove the bolts securing the transmission to the engine assembly and steering housing to transmission, pry up on the steering housing to clear the transmission housing; and roll the transmission and rear axle assembly to the rear out of the side frames.

SINGLE PLATE CLUTCHES

164. **ADJUSTMENT.** The accompanying table lists the tractor models with their transmision and clutch applications as well as the pedal free travel and pedal height measurements where height measurements are applicable. When adjusting pedal free travel refer to table for the figure number which will show the clutch linkage and points of adjustment.

Note: "Select-O-Speed" models are not listed in the table, see paragraph 240 for service.

NOTE: It is very important when renewing the clutch disc and/or pressure plate and cover assembly, that correctly matching

parts be used. The various assemblies can be identified as follows:

9-Inch Regular Clutch: (Early 4-speed Only) Clutch disc has solid hub (no cushioning springs). Use NAA-7550-A Disc Assembly with 8N-7563 Pressure Plate and Cover Assembly.

9-Inch Heavy Duty Clutch: (Early 4-speed Only) Clutch disc has cushioning springs in hub. Use 91A-7550 Disc Assembly with 09A-7563 Pressure Plate and Cover Assembly.

10-Inch Regular Clutch: (Late 4-speed and all 5-speed single clutch) Non-metallic facing material is used on disc and pressure plate, and cover assembly **does not** have adjusting screws on release fingers. Use NCA-7550-A Disc Assembly (10 splines in hub) or NDA-7550-B Disc Assembly (15 splines in hub) with NDA-7563-A Pressure Plate and Cover Assembly.

10-Inch Heavy Duty Clutch: (Late 4-speed and all 5-speed single clutch) Release fingers on pressure plate and cover assembly have adjusting screws. Use 313299 Disc Assembly (10 splines in hub) or CONN-7550-C Disc Assembly (15 splines in hub) with 311485 or B8NN-7563-A Pressure Plate and Cover Assembly.

10-Inch Extra Heavy Duty Clutch: (Late 4-speed and all 5-speed single clutch; can be used on models having 15 splines on clutch shaft only) Disc assembly has sintered metal facing, and pressure plate and cover assembly **does not** have adjusting screws on release fingers. Use B9NN-7550-A Disc Assembly only with B9NN-7563-A Pressure Plate and Cover Assembly. Is standard equipment on late production Heavy Duty Industrial models.

165. **R&R AND OVERHAUL.** To remove the clutch, it is first necessary to detach (split) engine from transmission as outlined in paragraph 161, 162 or 163. Then, depress the clutch release levers and insert wooden

CLUTCH PEDAL ADJUSTMENT TABLE

Transmission	Type	Pedal Free Play	Pedal Height	Refer to
Four-Speed	(Early)	¾-1	. . . .	Fig. FO127
Four-Speed	(Late)	1 ½	7	Fig. FO128
Four-Speed	(Offset)	1 ½	In line with brake pedal	Fig. FO129
Five-Speed Single		1 ½	7	Fig. FO128
Five-Speed Dual	(Early)	¾-1	10	Fig. FO131
Five-Speed Dual	(Late)	1 ½	9⅜	Fig. FO131

wedges between the release levers and cover plate. Unbolt clutch cover plate from flywheel and withdraw the cover assembly and lined friction disc.

Check release bearing and pilot bearing for damage or wear and renew if necessary. Also check clearance of clutch release shaft in its bushings which are located in transmission housing.

NOTE: A kit is available to install a pressure lubricated clutch release bearing (standard equipment in late industrial models) in all four speed transmission models and in five speed transmission models with single clutch after 1958-62 production Serial No. 14256.

Fig. FO127 — To adjust clutch pedal free play on early 4-speed, remove link pin and turn eye-bolt (A) in or out as required.

Fig. FO128 — On late four speed and five speed single clutch, first adjust free play by means of stop screw (A), then adjust pedal height at clevis (H).

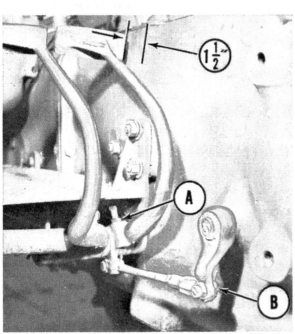

Fig. FO129 — To adjust clutch linkage on Offset tractor, first adjust pedal height by stop screw (A) then adjust pedal free play to 1½ inches by means of connecting link (B).

Components of the clutch cover and pressure plate assembly, except for adjusting screws on heavy duty 10-inch clutch, are not available for service. Friction facings are available separately from the clutch lined disc except for extra heavy duty clutch with sintered facings.

To adjust finger height on 10-inch heavy duty clutch, mount a new clutch disc with the cover assembly on the flywheel and place the adjusting fixture (See Fig. FO130) over the cover assembly so that the legs rest on the flywheel surface; then, adjust the screws until they just contact the surface (B) of fixture. Stake the screws in place after adjustment is complete.

Before installing clutch on flywheel, be sure that splines on clutch shaft and in clutch hub are clean and free of rust or corrosion. Then apply a very thin coat of light silicone grease (Ford specification No. M1C-43) to both splines. When installing the clutch, tighten the retaining cap screws to a torque of 12-16 Ft.-Lbs. and remove wedges from between release levers and cover plate.

DUAL CLUTCH

The Long 9 inch diameter clutch used on models with live pto is of the double plate, spring loaded, dry disc type. Depressing the clutch pedal about half way disengages the forward half of the dual clutch assembly and interrupts the power flow to the transmission. Depressing the pedal further disengages the rear half of the dual clutch assembly and interrupts the power flow to the power take-off input shaft.

If desired, the clutch pedal can be set for operation on the transmission (front) clutch only by shifting the clevis pin to the front hole in the clevis.

166. ADJUSTMENT. Desired clutch pedal free travel is ¾-1 inch as shown in Fig. FO131 for models prior to 1958. On later models, clutch pedal free travel is 1½-1¾ inches. If the pedal free travel is not as specified, loosen the lock nut (S) and turn the pedal free travel adjusting screw either way as required. Obtain the desired pedal height of 10 inches above the step plate for models prior to 1958; or 9⅝ inches for later models, by loosening the jam nut and turning clevis.

167. R&R AND OVERHAUL. To remove the clutch, it is first necessary to perform a tractor split between the engine and transmission housing as outlined in paragraph 161 or 162. Unbolt and withdraw the dual clutch assembly from flywheel.

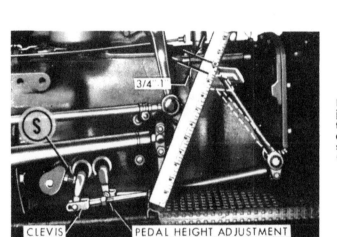

Fig. FO130 — Fixture for adjusting the heavy duty clutch fingers can be made locally by following the above diagram. Make from ⅛-inch stock and keep points A, B and C parallel.

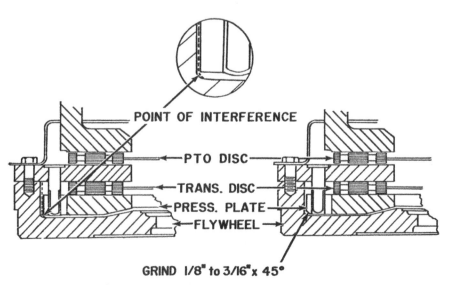

Fig. FO131 — To adjust clutch pedal free travel on models with a 5-speed transmission, loosen lock nut (S) and turn adjusting screw as required.

POINT OF INTERFERENCE

PTO DISC

TRANS. DISC

PRESS. PLATE

FLYWHEEL

GRIND 1/8" to 3/16" x 45°

Fig. FO132 — The dual clutches used on 1958 models can be used on earlier models providing the above modification is performed. Failure to make the modification will result in restricted travel and may prevent transmission and/or PTO clutch from being released.

Check the release bearing and pilot bearing for damage or wear and renew if necessary. Check also the running clearance of the clutch release shaft in its bushings which are located in the transmission housing.

NOTE: On late production tractors a heavier dual clutch assembly was incorporated. This newer clutch can be used on earlier models providing the following modification is performed.

Grind or file a ⅛"-3/16" x 45° chamfer on the forward edge of the pressure plate, being careful not to foul the drive blocks. Refer to Fig. FO132. This can best be accomplished by using a portable type grinder. Care should be taken to prevent filings from remaining in the clutch assembly.

Servicing procedures and specifications remain the same for both the early and late clutches.

Correlation mark the main drive plate, center pressure plate and the pto drive plate, as shown at (CM) in Fig. FO133, so the parts may be assembled in the same relative position. Using a valve spring compressor, compress the springs and remove the keepers, spring seats and springs. Remove the cap screws retaining cover to center plate and the three pins retaining the struts to the forward pressure plate and separate the units. Do not disassemble the pto pressure plate, as component parts are not catalogued. Refer to Fig. FO134.

Check the component parts in the conventional manner and renew any which are damaged or show wear. The spring retainers (keepers) should be discarded and new ones installed during assembly. Renew any spring that does not test 98-108 lbs. for the old (not ventilated) clutch; or 114-124 lbs. for the new (ventilated clutch), when compressed to a height of 1.671 inches.

Before reassembling clutch, be sure splines in friction disc hubs are clean and free of rust or corrosion. Then apply a thin film of light silicone grease (Ford specifiaction M1C-43) to splines.

The use of Nuday special aligning tools will facilitate reassembly as follows: Install the transmission pressure plate on fixture tool No. NDA-7502-1, then install the transmission lined disc (7550) so that longer hub of disc will be toward flywheel end of the clutch assembly as shown in Fig. FO135. Install the center drive plate, aligning the previously affixed assembly marks, making certain that the drive dogs properly engage the transmission pressure plate. Install the pto lined disc with hub of disc toward rear of clutch assembly as shown in Fig. FO136. If new plates are being installed note each assembly has a paint mark indicating heavy side. These marks should be located 120 degrees apart for correct balance.

Install aligning spacer tool No. NDA-7502-3 with taper toward flywheel side of the assembly as shown. Install the pto pressure plate and cover assembly, then install the strut pins and cotter pins. Install the release levers and tighten their retaining cap screws to a torque of 12-15 Ft.-Lbs. Install the pressure springs, seats and new retainers. Make certain that closed end of retainers are toward center of clutch; then, lock opposite end by pinching with needle nose pliers.

Install adjusting tool No. NDA-7502-5, flat washer and nut as shown in Fig. FO137 and tighten the nut until the three adjusting spacers No. NDA-7502 can be positioned against the lined discs between the pressure plates, 120 degrees apart as shown. Make certain that blocks marked top are toward the clutch springs. After making certain that the aligning spacer No. NDA-7502-3 bottoms on the pilot for the transmission lined disc, invert tool No. NDA-7502-5 and

Fig. FO134 — Component parts of the double plate clutch. Component parts of the PTO pressure plate are not catalogued.

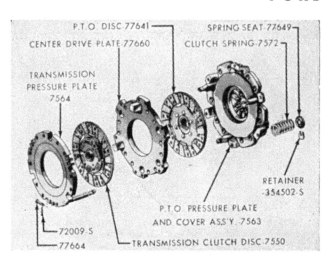

Fig. FO135 — View showing transmission clutch plate in position on fixture tool NDA 7502-1. Longer hub of disc is toward the flywheel.

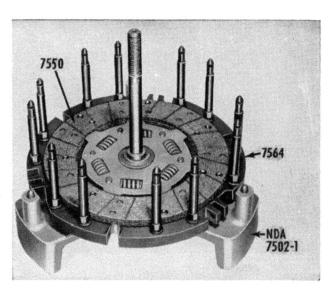

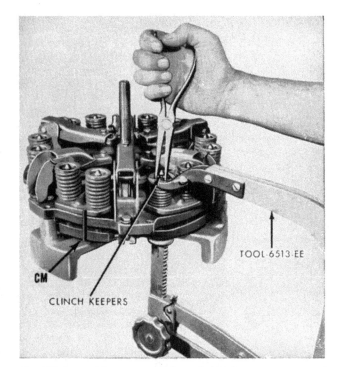

Fig. FO133 — Place correlation mark (CM) on components of double plate clutch so it may be assembled in the same relative position.

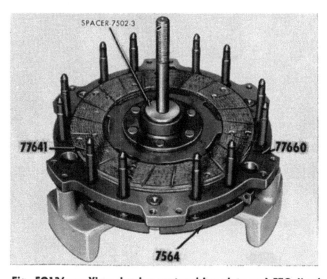

Fig. FO136 — View showing center drive plate and PTO lined disc in position. Hub of lined disc goes toward rear of clutch assembly as shown.

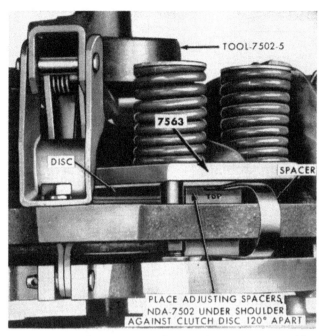

Fig. FO137 — View showing adjusting block in correct position to adjust clutch fingers on dual plate clutches.

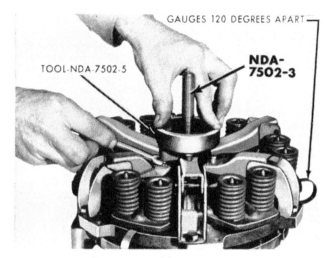

Fig. FO138 — Method of checking and adjusting clutch fingers. Refer to text for proper procedure.

reinstall it as shown in Fig. FO138. The highest machined surface of the tool is used to check and adjust the pto finger screws and the low side is used to check and adjust the transmission finger screws.

Now, while holding the adjusting tool firmly against the disc aligning spacer, turn each of the finger adjusting screws until there is a light drag on a 0.005 feeler gage held between the head of each adjusting screw and the machined surfaces of the adjusting tool. After the adjustment is completed and rechecked, invert the adjusting tool, install flat washer and nut and tighten the nut until the three adjusting spacers can be removed. Remove the adjusting tools 7502-3 and 7502-5.

Before reinstalling clutch, be sure splines on transmission and pto input shafts are clean and free of rust or corrosion. Then apply a thin coat of light silicone grease (Ford specification No. M1C-43) to splines. When installing the assembled clutch, tighten the retaining screws to a torque of 12-16 Ft.-Lbs.

COMBINATION TRANSMISSION

Tractors equipped with a four speed transmission have available a Combination (Step Up/Step Down) Transmission (formerly known as a Sherman Combination Transmission) as a factory or a field installation. This three range (over, under and direct) transmission is interposed between the engine clutch and the main transmission. When used in conjunction with the four-speed tractor transmission, twelve forward and three reverse speeds are possible.

REMOVE AND REINSTALL

170. Separate (split) tractor as outlined in paragraph 161, 162 or 163, then disconnect the clutch release bearing springs and remove the bearing. Remove the left brake pedal and the brake cross shaft Note: On later models (Ser. No. 14257 and up, of 1958-62 production) the brake cross shaft has been relocated under rear

axle center housing. Drain transmission as some models may have oil transfer holes between main and auxiliary gear case. See note following this paragraph. Remove the socket head screw from the combination transmission shift lever and remove the lever, then unbolt and remove the combination transmission from the mounting flange on front face of the main transmission housing.

The combination transmission mounting flange can be removed at this time by removing the four socket head screws and pulling flange and shims from the tractor transmission houing as shown in Fig. FO139.

Reinstall by reversing removal procedure and check bearing pre-load of the combination transmission as outlined in paragraph 175.

NOTE: On latest production units, a hole

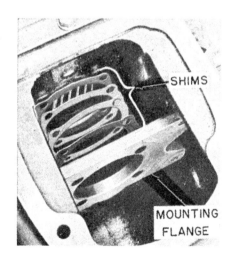

Fig. FO139—View showing mounting flange exploded from front wall of tractor transmission. Notice the shims between flange and transmission which control preload of transmission mainshaft and combination transmission output shaft bearings.

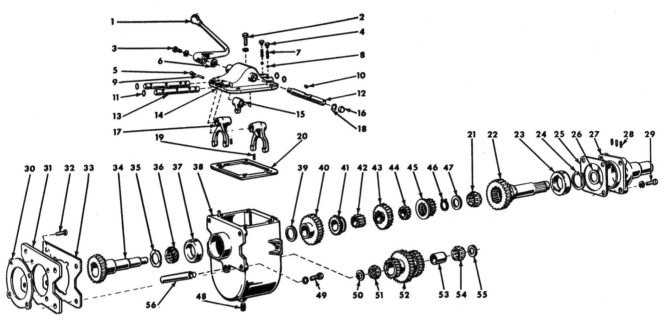

Fig. FO140 — Exploded view of Combination (Step Up/Step Down) Transmission that is available as optional equipment on some four-speed transmission models. Transmission can also be field installed by removing four-speed transmission clutch shaft and housing. Combination Transmission for tractors prior to serial No. 14257 of 1958-62 production has 10 splines on clutch shaft (22); later Combination Transmission has 15 splines on clutch shaft. Other components of both transmissions are similar.

1. Shift lever	11. Welch plugs	21. Roller bearing	31. Mounting flange	39. Washer	48. Drain plug
2. Cap screws	12. Selector shaft	22. Clutch shaft	32. Socket head screws	40. Step-down gear	49. Cap screws
3. Socket head screw	13. Shift rods	23. Ball bearing	33. Gasket	41. Shift collar	50. Thrust washer
4. Detent caps	14. Shift cover	24. Snap ring	34. Drive shaft	42. Splined sleeve (long)	51. Roller bearing
5. Set screw	15. Selector fork	25. Gasket	35. Oil slinger	43. Step-up gear	52. Cluster gear
6. Seal	16. Welch plug	26. Oil seal	36. Bearing cone &	44. Splined sleeve (short)	53. Bearing spacer
7. Detent springs	17. Shift forks	27. Housing	roller assy.	45. Shift collar	54. Roller bearing
8. Detent balls	18. Snap ring	28. Set screws	37. Bearing cup	46. Snap ring	55. Thrust washer
9. Interlock	19. Set screws	29. Cap screws	38. Gear case	47. Thrust washer	56. Cluster shaft
10. Woodruff key	20. Cover gasket	30. Shims			

is provided in the mounting flange and shims so that lubrication oil will be maintained at same level in the main and auxiliary transmission gear case. When reinstalling mounting flange and shims, be sure the oil holes are aligned.

OVERHAUL

171. SHIFTER RAILS AND FORKS. To remove shifter rails and forks, the combination transmission must be removed as outlined in paragraph 170. Refer to Fig. FO140, remove the retaining cap screws and lift shifter cap assembly (14) from gear case. Remove the two detent caps (4), detent springs and detent balls from shifter cap. Remove lock wires and the drilled head set screws from shifter fingers (forks) (17). Remove the Welch plugs (11) from the shifter rod bores at the rear of the shifter cap, then drive the shifter rods (13) out the front of shifter cap. Remove the set screw from the interlock bore and remove the interlock (9). Remove the snap ring from the groove on the selector shaft (12) and slide the selector fork (15) and snap ring (18)

toward the end of the selector shaft which is opposite from the shoulder stop. Remove the Woodruff key (10) from selector shaft then withdraw selector shaft from the shifter cap (14). Remove selector shaft seal (6) from bore in shifter cap.

Wash all parts in a suitable solvent and proceed as follows: Position the shifter rods (13) and selector shaft (12) in their respective bores and check for freedom of movement; if binding occurs due to rods or shaft being bent, renew the bent part. If binding occurs due to rods or shaft being scored, it may be possible to recondition the scored part by using Crocus cloth. Inspect the shifter fingers (forks) (17) and selector fork (15) and renew them if they show signs of contact at points other than the contact pads. Inspect the interlock (9) for flat spots or signs of scoring and renew if necessary.

172. When reassembling use new seal (6) and Welch plugs and proceed as follows: Place the selector shaft oil seal (6) in its bore and drift into position using a socket of proper size.

Start the selector shaft in its bore and place the selector fork (15) and snap ring (18) on shaft. Install Woodruff key in its slot then slide selector shaft (12) into place and position snap ring in its groove. Start a shifter rod (13) in the bore farthest (opposite) from the interlock plug and be sure the end with the two grooves is toward the detent end of shifter cap. Place a shifter finger (fork) (17) over shifter rod so that the recess in shifter finger will engage selector fork (15). Note: The shifter fingers (forks) and shifter rods are identical and can be interchanged. Position parts and align the center groove in the shifter rod (13) with the tapped hole in the shifter finger (17), then secure shifter finger with the drilled head cap screw (19). Install the interlock and tighten socket head screw (5). Install the other shifter rod (13) and shifter finger (17) in the same manner as the first was installed, then install lockwires in the drilled head set screws and the holes provided in the shifter fingers (17). Place the detent balls (8) and springs (7) in their bores and install the detent caps (4). Install new Welch plugs.

Fig. FO141 — View showing clutch shaft and support assembly removed from gear case. Note thrust washer.

Fig. FO142 — Clutch shaft and bearing assembly can be removed from front support after loosening Allen screws in support.

Fig. FO143 — View showing drive (output) shaft partially disassembled. Note the position of the long and short spline sleeves.

Use a new gasket and install shifter cap assembly to gear case and make sure the shift fingers engage the shift collars. Reinstall combination transmission to tractor transmission as in paragraph 170 and rejoin tractor as in paragraphs 161, 162 or 163.

173. CLUTCH SHAFT. With tractor separated (split) as outlined in paragraphs 161, 162 or 163, and the unit drained, the clutch shaft (22—Fig. FO140) can be removed by unbolting the support assembly (27) and pulling the support assembly and clutch shaft from the gear case. Remove the roller bearing (21) and the thrust washer (47) from the pilot end of drive shaft (34). Remove the three socket head screws (28) from front support (27) and remove the clutch shaft assembly from support. Remove the snap ring (24) and press ball

bearing (23) from clutch shaft. Remove oil seal (26) from front support using an Owatonna No. 956 bearing puller and slide hammer, or its equivalent.

Wash all parts in a suitable solvent and check the ball bearing (23) and roller bearing (21) for rough spots, flat spots, freedom of movement or other signs of wear. Check clutch shaft (22) for chipped teeth, cupping of the pilot bore, scoring or other signs of wear or damage. Check thrust washer (47) for scoring or undue wear. Renew parts as needed.

When reassembling use new oil seal (26) in front support (27). Press bearing (23) on clutch shaft and secure with snap ring (24). Install the clutch shaft and bearing in the front support and secure with the three socket head screws (28). Note: Use caution when inserting shaft into support to avoid damaging oil seal. Fill unit with proper lubricant. Place the thrust washer (47), then the roller bearing (21) on the pilot end of drive shaft (34) and using a new gasket install the front support and clutch shaft assembly after making certain that the oil return hole in front support is on the bottom.

With unit reassembled, rejoin engine and transmission housing.

174. DRIVE SHAFT. With the combination transmission removed as outlined in paragraph 170, remove shifter cap assembly as in paragraph 171 and the front support and clutch shaft assembly as in paragraph 173. Refer to Fig. FO140 and remove snap ring (46) from drive shaft (34); then remove front shift collar (45), short spline sleeve (44), step-up gear (43), rear shift collar (41), long spline sleeve (42), step-down gear (40) and washer (39) from drive shaft (34). NOTE: Identify shift collars (41 & 45) so they can be reinstalled in their proper positions. Pull the drive shaft (34), oil slinger (35) and taper bearing (36) from rear of gear case; then, using a suitable press or puller, remove the taper bearing and oil slinger from the drive shaft. Remove bearing cup (37) from gear case, if necessary.

Clean all parts in a suitable solvent and inspect the taper bearing (36) for rough spots, freedom of movement or other signs of wear. Inspect all drive shaft gears for erratic wear patterns, damaged teeth or splines, cupping, overheating on the gear end faces or scoring of inside diameters and renew as necessary.

Reinstall drive shaft as follows: Place oil slinger (35) on drive shaft and then press on the taper bearing. Place shaft and bearing in gear case and install parts as follows: Install gear washer (39) and step-down gear (40) with hub of gear facing rearward. Install long spline sleeve (42), then position the rear shift collar (41) with teeth toward rear. Install step-up gear (43) with the shift collar engaging teeth rearward. Install the short spline sleeve (44), then position the front shift collar (45) with engaging teeth toward front. Now install snap ring (46). Install the shifter cap assembly.

Note: With the drive shaft assembly and the shifter cap assembly installed, a bearing pre-load must be established in the Combination Transmission as follows:

175. Remove mounting flange (33) and install approximately 0.050 of shims (30) behind the mounting flange as shown in Fig. FO139. Be sure flat side of mounting flange is at top and reinstall the flange. Shims are available in thicknesses of 0.007 and 0.012. Position new mounting flange gasket and install the combination transmission; at which time, the drive shaft should have some end play. Now, remove shims (30—Fig. FO140) from behind mounting flange until drive shaft has zero end play, then remove one 0.007 thick shim to pre-load the drive shaft bearing.

176. Fill unit with specified lubricant and install the clutch shaft assembly as outlined in paragraph 173; then, rejoin engine and transmission housing.

177. CLUSTER GEAR AND SHAFT. Remove the combination transmission from tractor as in paragraph 170 and the shifter cap, clutch shaft and drive shaft assemblies as outlined in paragraphs 171 through 174. Refer to Fig. FO140 and drift cluster shaft (56) toward rear of gear case and catch cluster gear (52) and thrust washers (50 & 55) as shaft is driven out. Remove the roller bearings (51 & 54) and the spacer (53) from the inside diameter of cluster gear (52).

Check cluster gear for chipped teeth, cupping of end faces, signs of overheating and wear of inside diameter. Check roller bearings for flat spots and freedom of movement or other signs of wear. Check cluster shaft for pitting, scoring or wear of outside diameter. If wear or damage is found on any of the above parts, it is recommended by the Ford Motor

Company that the cluster shaft, gear and bearings be renewed as an assembly.

Reassembly is evident; however, keep in mind that the larger of the two end gears of the cluster gear is toward the front of the gear case. Reassemble the balance of the transmission as outlined in paragraphs 171 through 174 and reinstall in tractor as outlined in paragraph 170.

REVERSING TRANSMISSION

Industrial tractors equipped with the four speed transmission have available as a factory installed option a Reversing Transmission. Field installation may be made in any Ford tractor equipped with the four speed transmission. The reversing transmission reverses the power flow forward of the main transmission, enabling the tractor to move forward or backward in any of the regular gear speeds. If field installation is made where tractor will be used in 4th gear reversed, also install special 4th gear coupler (B9NN-7106-A) to prevent transmission from jumping out of gear.

REMOVE AND REINSTALL

178. Separate (split) the tractor at the clutch housing as outlined in paragraph 161, 162 or 163, disconnect the clutch release bearing springs and remove the bearing. Remove the socket head cap screw securing the shift lever pivot arm to the transmission and disconnect the pivot arm; then, unbolt and remove the reversing transmission from the mounting flange on front face of main transmission housing.

The reversing transmission mounting flange can be removed at this time by removing the four socket head mounting screws and withdrawing flange and shims from the tractor transmission housing as shown in Fig. FO139.

Reinstall by reversing the removal procedure after first checking the bearing pre-load as outlined in paragraph 179.

OVERHAUL

179. To overhaul the reversing transmission after it has been removed, first unbolt and remove the input shaft housing (38—Fig. FO144) together with the shaft (24) and bear-

ing (34). Shaft and bearing can be removed from the input housing after removing snap ring (23). Seal (36), bearing (34) and pilot bearing (22) can be renewed at this time. Unbolt and remove the shifter cap assembly (2) and drift the shifter rail (11) out of the housing and shifter fork assembly. Two detent balls and one spring are contained in a drilling in the shifter rail and engage in detents in the shifter fork. Lift the shifter fork out of the housing, remove thrust washer (21) and snap ring (20) from the front end of the main drive gear shaft (29) and slide the shaft and bearing (31) rearward out of the housing. Remove in turn the shift collar (18), sleeve (19), reverse gear (15) and spacer (13) as the main gear and shaft are withdrawn. Main drive gear (29), bearing (31), and the needle bearings (14 and 16) in the reverse drive gear can be renewed at this time if required. Drive the reverse idler shaft (40) forward out of the housing and remove the gear (43) and thrust washers (41 and 46) from the housing. Drift the cluster gear shaft (50) rearward out of the housing and remove the cluster gear (53) and spacers from the housing.

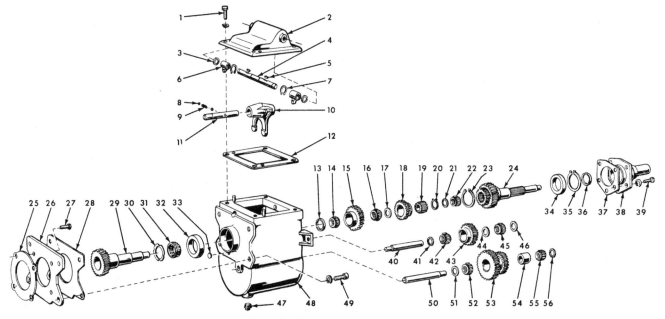

Fig. FO144 — Exploded view of reversing transmission which is optional factory equipment on some models with four-speed transmission and is also available for field installation. When servicing transmission, it is recommended that new type non-directional slinger (30), Ford part No. CONN-7C237-B, be installed in place of slinger used on earlier models.

1. Cap screws	11. Shift rail	21. Thrust washer	31. Bearing cone & roller assy.	39. Cap screws	48. Gear case
2. Shift cover	12. Cover gasket	22. Roller bearing	32. Bearing cup	40. Idler shaft	49. Cap screws
3. "O" ring	13. Spacer	23. Snap ring	33. Expansion plug	41. Thrust washer	50. Cluster shaft
4. Selector shaft	14. Roller bearing	24. Clutch shaft	34. Bearing	42. Roller bearing	51. Thrust washer
5. Woodruff keys	15. Reverse gear	25. Shims	35. Snap ring	43. Idler gear	52. Roller bearing
6. Selector forks	16. Roller bearing	26. Mounting flange	36. Oil seal	44. Spacer	53. Cluster gear
7. Snap rings	17. Thrust washer	27. Socket head screws	37. Gasket	45. Roller bearing	54. Spacer
8. Steel ball	18. Shift collar	28. Gasket	38. Housing	46. Thrust washer	55. Roller bearing
9. Spring	19. Splined sleeve	29. Drive gear		47. Drain plug	56. Thrust washer
10. Shift fork	20. Snap ring	30. Oil slinger			

NOTE: A new non-directional oil slinger, Ford part No. CONN-7C237-B, is now used to prevent oil from being pumped out of the reversing transmission into the four-speed transmission during prolonged operation with the drive shaft (29—Fig. FO144) turning counter-clockwise. To install this slinger, remove bearing (31) and install new slinger in place of original slinger (30). Be sure face of new slinger having raised ribs is placed towards bearing (31).

Clean all of the parts in a suitable solvent, examine them for wear or damage and renew as required. Assemble the transmission by reversing the disassembly procedure but do not install the input gear and housing at this time.

Install the transmission, less the input shaft, in the tractor and check the transmission main shaft pre-load. Main shaft pre-load is adjusted by means of shims located between the mounting plate (26) and the main transmission housing, as shown in Fig. FO139. The correct pre-load of 0.005-0.007 is established by removing shims until shaft has zero end play, then removing one additional 0.007 shim. After correct pre-load has been established, add one quart of oil to the reversing transmission through the front opening, make sure thrust washer (21—Fig. FO144) is in place and install the input shaft and housing.

TRANSMISSION

(Four Speed)

REMOVE AND REINSTALL

185. To completely overhaul the transmission, it is first necessary to remove the unit as follows: Drain transmission, hydraulic system and rear axle center housing and remove power take-off shaft. Split engine from transmission as outlined in paragraph 161, 162 or 163.

If equipped with a combination or reversing transmission, remove shift lever, then unbolt unit from front face of transmission housing.

186. Remove power take-off shifter control from left side of rear axle center housing; then, remove the step plates, brake and clutch linkage. Block up the rear axle center housing and unbolt same from transmission. Separate transmission and rear section and place the transmission on bench or stand.

OVERHAUL

187. **SHIFTER RAILS AND FORKS.** Rails and forks may be removed without splitting the transmission from the engine, but the transmission top cover must be removed and the rear axle center housing must be separated from transmission case as outlined in paragraph 186. Lift out the top shifter rail spring (25—Fig. FO-146) and detent ball, back off shifter fork lock screw (22) and remove shifter rail and fork. Remove shift plate pivot screws (10) from both sides of transmission housing and remove shift plates.

188. Remove both of the lower shift rails and forks through top opening after removing pivots (10) and the detent springs and balls.

Renew worn or damaged parts and reinstall. The square cornered slots in lower rails for shift plates must face inward.

NOTE: Effective at tractor Serial Number 14257 of 1958-62 production, plugs have been installed in the center housing end of transmission case at the lower shift rail positions. Be sure plugs are installed and in good condition. Plugs were installed to eliminate the possibility of oil leaking into the hydraulic system reservoir.

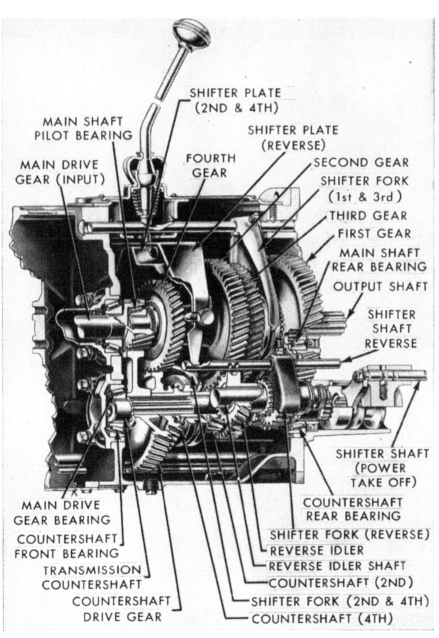

Fig. FO145 — Cutaway side view of four-speed transmission.

189. MAIN DRIVE GEAR. (Clutch Shaft on models without combination or reversing transmission.) This input shaft (Fig. FO145) may be removed after transmission and engine are separated as in paragraph 161, 162 or 163. However, if adjustment of clutch shaft bearings is required, it will be necessary to perform the rear section split (as in paragraph 186) to gain access to the mainshaft bearing shims which control the adjustment. Remove input shaft and retainer (1 and 6—Fig. FO148) as a unit.

Renew worn or damaged parts and reassemble. Coat shaft with oil and rotate same when assembling it to the retainer so as to avoid damaging the oil seal. Lip of seal should face toward rear. After shaft is reinstalled, adjust end play as per paragraph 191. However, if the mainshaft is not being overhauled, end play of 0.000-0.002 can be tolerated. Tighten the main drive gear bearing retainer cap screws to a torque of 25-30 Ft.-Lbs.

190. MAIN (OUTPUT) SHAFT. To remove this shaft (Fig. FO145), first remove transmission from tractor as per paragraphs 185 and 186. Then, remove the top shifter rail and fork and shift plates as per paragraph 187.

Remove the mainshaft rear bearing retainer (21—Fig. FO148) and the adjusting shims. Inspect the output shaft seal (22) and renew if necessary. Remove the clutch shaft (main drive gear) and retainer as a unit; or the combination or reversing transmission, if so equipped. Remove the reverse idler shaft shift rail, allowing fork to remain in housing. Lift out the mainshaft and gear cluster as a unit. Disassemble the unit and renew any worn or damaged parts. Refer to Fig. FO148 as a guide for reassembly.

Before installing the main shaft, it is advisable to check the countershaft preload as shown in Fig. FO149. Desired preload of 15-30 inch lbs. is controlled by shims (36—Fig. FO148) located between the pto support and rear face of transmission.

Reinstall the assembly to the transmission by reversing the removal procedure. Move the shifter forks to neutral position and check the adjustment of the main shaft bearings. If a scale is not available, adjust bearings to a very slight preload. If a scale is available, check and adjust the bearings as per paragraph 191.

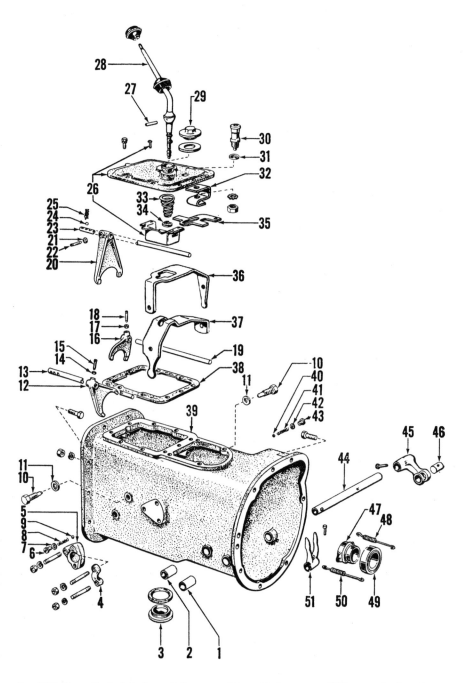

Fig. FO146 — Exploded view of four-speed transmission gear shifting mechanism, transmission housing and related parts. Reverse shifter fork (16) is different than shown above; refer to Fig. FO145.

1. Clutch shaft bushing	15. Lock screw	28. Shift lever	40. Detent ball
2. Brake shaft bushing	16. Reverse shift fork	29. Filler plug	41. Spring
3. Drain plug	17. Lock nut	30. Starter switch	42. Gasket
4. Radius rod cap	18. Lock screw	31. Gasket	43. Detent spring seat
5. Radius rod socket	19. Reverse shift rail	32. Starter switch	44. Clutch shaft
6. Detent spring seat	20. 1st & 3rd shift fork	support	45. Clutch lever
7. Gasket	21. Lock nut	33. Spring	46. Pin
8. Spring	22. Lock screw	34. Spring seat	47. Clutch bearing hub
9. Detent ball	23. 1st &— 3rd shift rail	35. Starter safety latch	48. Clutch release
10. Shift plate pivot	24. Detent ball	36. Reverse shift plate	spring
11. Gasket	25. Spring	37. 2nd & 4th shift plate	49. Clutch release
12. 2nd & 4th shift fork	26. Shift cover assy.	38. Gasket	bearing
13. 2nd & 4th shift rail	27. Pin, shift lever	39. Transmission	51. Clutch release fork
14. Lock nut		housing	

Fig. FO147 — Desired adjustment of main shaft bearings is 20-35 inch-pounds torque to rotate the shaft.

191. To test bearing adjustment, rotate rear end of main shaft with transmission in neutral, and measure turning torque as shown in Fig. FO-147. If torque is 20 to 35 inch pounds measured with shaft in motion, bearing adjustment is correct. Vary the number of shims under rear bearing retainer to obtain this adjustment.

Note: If the countershaft preload is on the low side, the main shaft preload should be on the low side. The reverse is true also. Tighten the rear bearing housing retaining cap screws to a torque of 25-30 Ft.-Lbs.

192. **COUNTERSHAFT.** This shaft (Fig. FO145) can be removed after mainshaft is out. Remove power take-off shifter and bearing support (38— Fig. FO148) and remove front bearing carrier from opposite end of transmission. Remove countershaft and gears as a unit through cover opening in top of transmission housing.

Renew worn or damaged parts, including the oil seal in the pto shifter support and reassemble as shown in Fig. FO148. Reinstall the assembly and adjust bearings by means of shims (36) to obtain zero end play and slight preload of 15-30 inch pounds torque to rotate shaft. Shaft should be rotated for checking bearing ad-

Fig. FO148 — Exploded view of gears and shafts used in four-speed transmission.

1. Input shaft housing
2. Gasket
3. Shaft seal
4. Bearing cup
5. Bearing cone
6. Input shaft
7. Pilot bearing
8. Washer
9. Mainshaft 4th gear
10. Mainshaft 2nd gear
11. Mainshaft 3rd gear
12. Connector
13. 1st/3rd coupling
14. Mainshaft 1st gear
15. Thrust washer
16. Rear bearing
17. Oil seal sleeve
18. Main shaft
19. Bearing cup
20. Shim pack
21. Bearing retainer
22. Oil seal
23. Countershaft front bearing retainer
24. Gasket
25. Bearing cup
26. Countershaft front bearing
27. Countershaft drive gear
28. Countershaft fourth gear
29. 2nd/4th coupling
30. Connector
31. Countershaft 2nd gear
32. Countershaft
33. Rear bearing
34. Oil sleeve
35. Bearing cup
36. Shim pack
37. Oil seal
38. Pto shifter support
39. Reverse idler
40. Reverse idler gear
41. Thrust washer
42. Thrust washer
43. Reverse idler shaft
44. "O" ring
45. Retainer

justment by inserting pto shaft in shifter unit as shown in Fig. FO149.

193. **REVERSE IDLER GEAR.** This gear (Fig. FO145) can be removed after mainshaft and countershaft are out. Pull idler shaft out toward rear and lift out gears. Rebush gears or renew parts if necessary, and reinstall, assembling as shown in Fig. FO148. The I&T suggested running clearance of the split and oil grooved bushings on idler shaft is 0.0025-0.004.

194. **PTO SHIFTER BEARING SUPPORT.** For information on overhaul of the pto shifter and bearing support, refer to paragraph 290.

Fig. FO149 — Desired adjustment of countershaft bearings is 15-30 inch-pounds to rotate the shaft when main shaft is out of the transmission.

TRANSMISSION
(Five Speed)

REMOVE AND REINSTALL

195. To completely overhaul the transmission it is necessary to remove the unit as follows: Drain the transmission, hydraulic system and rear axle center housing and remove the power take-off shaft. Split engine from transmission as outlined in paragraph 161 or 162.

196. Remove the power take-off control from left side of rear axle center housing; then, remove the step plates, brake linkage (on models prior to Serial Number 14257 of 1958-62 production) and clutch linkage. Block up the rear axle center housing and unbolt same from transmission. Separate

tranmission and place transmission on bench or stand.

OVERHAUL

197. **SHIFTER RAILS AND FORKS.** Rails and forks cannot be removed until cluster gear (main) shaft has been removed as outlined in paragraph 201.

198. Remainder of the procedure is as follows: Refer to Fig. FO151. Remove the three shift rail plugs (42), springs (44) and plungers (45) from left side of transmission case.

Drive out the three Welch plugs from rear of case behind the shift rails. Remove set screw from third and reverse speed shift fork (36) and gate (23). Remove third and reverse (top) rail (39) out through Welch plug hole. Keep rail, fork and gate together.

To remove the first and second gear (middle) rail (40), shift fork (15) and gate (21), remove set screws from same and withdraw rail through rear opening. The fourth and fifth speed (bottom) rail (41) and combined fork and gate (16) can be similarly removed after removing the set screw from the combination fork and gate (16). If interlock plate (22) and its guide rod (20) are to be removed do so by withdrawing rod through bolt hole in top of case or rod can be driven out towards bottom of transmission after removing the hexagon head plug from bottom of transmission case. Interlock is not reversible, be sure to install so that it is to the outside of guide rod when parallel with housing.

To remove gear shift shaft (37) remove the detent plug, spring and plunger from the left side of the

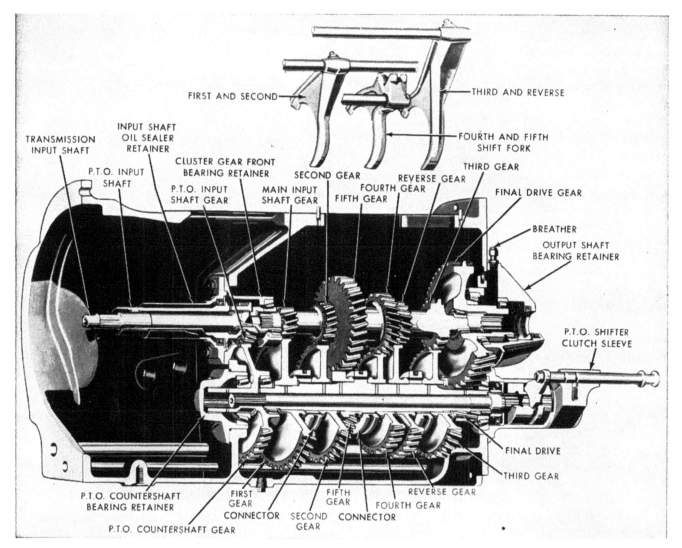

Fig. FO150 — Cutaway view of five-speed transmission and associated pto drive parts. The hollow pto input shaft is used only on models with live pto.

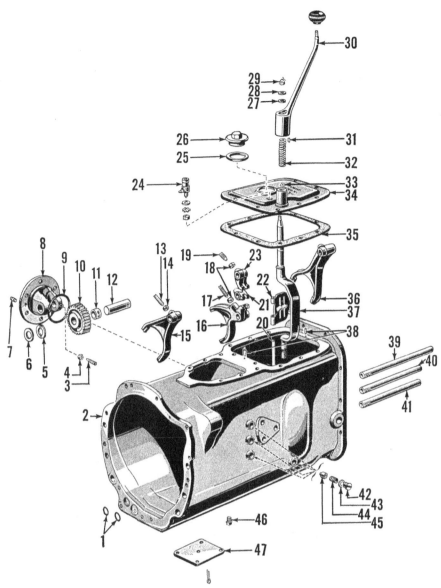

FO151 — Exploded view of five-speed transmission case, shifter mechanism and reverse idler assembly.

Fig. FO152 — Preload of five-speed transmission gear shafts should be checked with a pull scale.

1. "O" rings	14. Lock nut	24. Starter switch	37. Shifter shaft
2. Transmission case	15. Shifter fork,	25. Gasket	38. Dowel pins
3. Set screw	1st & 2nd	26. Filler cap	39. Shift rail,
4. Lock nut	16. Shifter fork,	27. Flat washer	3rd & Rev.
5. Thrust washer	4th & 5th	28. Lock washer	40. Shift rail,
6. Thrust washer	17. Set screw	29. Acorn nut	1st & 2nd
7. Dowel pin	18. Lock nuts	30. Shift lever	41. Shift rail,
8. Reverse idler	19. Set screw	31. Woodruff key	4th & 5th
bracket	20. Guide rod for (22)	32. Coil spring	42. Detent plugs
9. "O" ring	21. Shifter gate,	33. Oil seal	43. Gaskets
10. Reverse idler	1st & 2nd	34. Shifter cover	44. Detent springs
11. Bushing	22. Interlock plate	35. Gasket	45. Detent plungers
12. Idler shaft	23. Shifter gate,	36. Shifter fork,	46. Drain plug
13. Set screw	3rd & Rev.	3rd & Rev.	47. Inspection hole
			cover

transmission housing and lift out the shaft.

199. REVERSE IDLER GEAR. Reverse idler gear (15—Fig. FO151) complete with bracket (8) can be removed at any time without disturbing any other parts by simply removing the housing retaining stud nuts. The renewable bushing (11) is available separately or as an assembly with the idler gear (10). After in-

stalling bushing in idler gear, ream bushing to provide 0.003-0.005 clearance between bushing and shaft (12). Be sure shaft set screw is to the rear when reinstalling idler bracket assembly.

200. CLUTCH SHAFT (TRANSMISSION AND PTO INPUT SHAFT). On models with dual clutch ("live" PTO), the transmission input shaft is surrounded by a hollow PTO input

shaft as shown in Fig. FO150. On models with single clutch, the transmission input shaft and PTO input shaft are integral. The input shaft(s) can be removed after the engine is separated from the transmission as outlined in paragraph 161 or 162. Remove either type shaft as follows:

Disengage clutch shaft spring from transmission housing, pull clutch fork forward and remove clutch throwout bearing from oil seal retainer (1 or 14—Fig. FO153). Remove the clutch throwout shaft from transmission housing and, on early production models with brake cross shaft in transmission housing, remove brake cross shaft also. Unbolt and remove the oil seal retainer and transmission input shaft(s). Single clutch input shaft (18) is fitted with one thrust washer (17) between PTO drive gear and oil seal retainer (14). Early double clutch models (with brake cross shaft in transmission case) use one thrust washer (5) between PTO drive gear on hollow input shaft (6) and oil seal retainer (1). Later production double clutch models (with brake cross shaft under center housing) are fitted with two thrust washers at (5).

Oil seal (2), needle bearing (3) and bushing (9) located in bore of hollow PTO input shaft (6) on double clutch models can be renewed at this time. Ream bushing after installation to provide 0.003-0.005 clearance between bushing and journal on transmission input shaft. After removing seal and needle bearing, pack area behind needle bearing location with wheel bearing grease. Press on lettered side of needle bearing cage until front side of cage is 0.76 inch behind front end of shaft. Pack area in front of needle bearing with wheel bearing grease

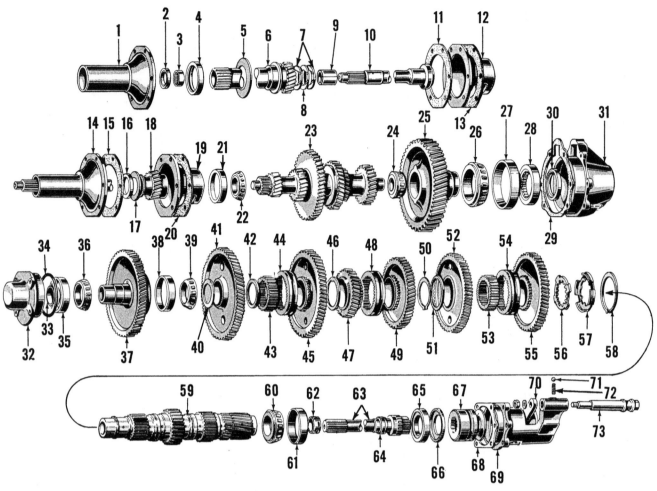

Fig. FO153 — Exploded view of five-speed transmission gears and related parts. Items 1 through 13 are used on live pto models only; items 14 through 20 are used on single clutch models only.

1. Oil seal retainer	12. Bearing retainer	24. Cone & roller assy.	37. PTO driven gear	50. Snap ring	62. Bushing
2. Oil seal	13. Gasket	25. Output gear	38. Bearing cup	51. Thrust washer	63. PTO output shaft
3. Needle bearing	14. Oil seal retainer	26. Cone & roller assy.	39. Cone & roller assy.	52. Reverse gear	64. Sleeve
4. Oil seal	15. Gasket	27. Bearing cup	40. Thrust washer	53. Splined connector	65. Oil seal
5. Washer	16. Oil seal	28. Roller bearing	41. First speed gear	54. Splined coupling	66. Retainer
6. PTO input shaft	17. Thrust washer	29. Shims	42. Thrust washer	55. Third speed gear	67. Coupling
7. Thrust washers	18. Input shaft	30. Oil seal	43. Splined connector	56. Thrust washer	68. Shims
8. Thrust washer	19. Bearing retainer	31. Retainer	44. Splined coupling	57. Retainer	69. Support
9. Bushing	20. Gasket	32. Bearing retainer	45. Second speed gear	58. Snap ring	70. Shifter stop
10. Transmission input shaft	21. Bearing cup	33. Bushing	46. Thrust washer	59. Countershaft	71. Detent ball
11. Gasket	22. Cone & roller assy.	34. "O" ring	47. Fifth speed gear	60. Cone & roller assy.	72. Detent spring
	23. Main cluster gear	35. Bearing cup	48. Splined coupling	61. Bearing cup	73. Shifter rail, PTO
		36. Cone & roller assy.	49. Fourth speed gear		

Install new seal with lip of seal towards needle bearing and lubricate seal with Lubriplate or similar grease. Three thrust washers are used between PTO input shaft (6) and shoulder on transmission input shaft (10). Install two outer washers (7) with oil groove side away from center washer (8) which has oil grooves on each side.

Oil seal (4 or 16) in retainer (1 or 14) can be renewed at this time. Install new seal with lip towards rear (flanged) end of retainer.

When reassembling, hook the inner end of the clutch shaft coil spring in clutch throwout fork; then, using a small box end wrench as a lever, hook outer end of spring in position on transmission case.

NOTE: Rust or corrosion of splines on transmission input shaft(s) which mate with splines in clutch disc hub is sometimes encountered. This condition can cause rapid wear of splines and/or improper operation of clutch. Before reassembling tractor, it is advisable to clean splines in clutch hub and on shaft of all rust or corrosion and apply a very thin coat of light silicone grease (Ford Specification M1C-43) to splines in hub and on shaft.

201. TRANSMISSION MAIN SHAFT (CLUSTER GEAR AND OUTPUT GEAR). The transmission cluster gear (23—Fig. FO153) and output gear (25) can be removed from the transmission after removing transmission from tractor as outlined in paragraphs 195 and 196. To remove cluster gear and output gear, proceed as follows:

Loosen acorn nut (29—Fig. FO151) on top of shift lever (30). While pulling upward on shift lever, strike acorn nut sharply with hammer to loosen lever from taper on shaft (37). Remove acorn nut, washers, and shift lever and spring (32); then, remove Woodruff key (31) from shaft (37). Cover (34) can now be unbolted and removed from transmission.

Remove reverse idler as outlined in paragraph 199, and remove oil seal retainer and transmission input shaft(s) as outlined in paragraph 200. Unbolt and remove bearing retainer (31) and shim pack (29) from rear end of transmission case. Support cluster gear and remove front bearing retainer (12 or 20). Lift cluster gear up and forward to clear countershaft second gear (45) and rear end of cluster gear is free from output gear; then, lift cluster gear from transmission rear end first. The output gear can now be removed from transmission by moving the top shifter rail forward far as possible and tipping gear forward to clear rear opening in case.

202. Bearing cone and roller assemblies (22 and 24) can be removed from cluster gear with OTC knife edge bearing puller attachment or similar tools. Cup for rear bearing cone and roller assembly (24) is integral with output gear (25). Access holes are provided in output gear through which a long, thin punch can be used to drive bearing cone and roller assembly (26) from rear side of gear. Bearing cup (27), roller bearing (28) and seal (30) can be removed from front side of retainer (31) with drift or punch inserted through rear opening. Install new seal (30) with lip towards front side of retatiner. Bearing cup (21) in front bearing retainer (12 or 19) can also be renewed at this time.

NOTE: In early 600 and 800 series tractors (prior to serial No. 21080), fifth speed gear on gear cluster (23—Fig. FO153) had 47 teeth. If necessary to renew this type cluster gear with newer cluster gear having 43 tooth fifth speed gear (largest gear on cluster), it will be necessary to renew the countershaft fifth speed gear (47) also. Early countershaft fifth speed gear had 33 teeth; late gear has 37 teeth.

203. Before reinstalling the cluster gear and output gear, the bearing adjustment (preload) of the countershaft bearings should be checked and adjusted if necessary as outlined in paragraph 207.

Reinstall cluster gear and output gear by reversing procedure outlined in paragraph 201. Before reinstalling top cover, check bearing adjustment of cluster gear (transmission main shaft) as follows: Wrap a cord around the cluster gear in position shown in Fig. FO152 and attach a pull scale to the cord. Bearing adjustment (preload) is correct if a steady pull of 8-10 lbs. on double clutch models or 10-12 lbs. on single clutch models is required to rotate the cluster gear.

NOTE: Be sure all sliding splined couplings (44, 48 and 54) are in neutral (centered) position when checking adjustment.

If excessive force is required to rotate the cluster gear, remove rear bearing retainer and increase thickness of shim stack (29). If force required to rotate the cluster gear is less than specified, decrease thickness of shim stack. Shims are available in thicknesses of 0.003, 0.005 and 0.012.

204. **TRANSMISSION COUNTERSHAFT ASSEMBLY.** To remove the countershaft assembly, first remove the cluster gear and output gear as outlined in paragraph 201 and remove shifter rails and forks as outlined in paragraph 198. Then, proceed as follows:

Remove the PTO support (69—Fig. FO153) and shifter assembly and shim stack (68) from rear end of transmission. Withdraw the PTO output shaft (63) from rear of countershaft. Remove the front bearing retainer (32) and "O" ring (34). Move the countershaft and gear assembly rearward until front end of shaft clears PTO driven gear (37) and lift the countershaft assembly out of the transmission front end first. Then, remove PTO driven gear from transmission.

205. Disassemble countershaft and gears as follows: Remove bearing cone and roller assembly (60) from rear end of countershaft. Remove snap ring (58), retainer (57) and washer (56). Slide third speed gear (55), coupling (54), connector (53), reverse gear (52) and thrust washer (51) from rear end of shaft. Then, remove snap ring (50) and slide fourth speed gear (49) and coupling (48) from rear end of shaft.

Using a gear puller or press, remove first speed gear (41), thrust washer (40) and bearing cone and roller assembly from front end of shaft. As shaft is hollow, a centering plug of correct size must be used. Thrust washer (42), connector (43), coupling (44), second speed gear (45), thrust washer (46) and fifth speed gear can then be removed from front end of shaft.

Bushing (62) for pto shaft (63) in rear end of countershaft (59) is renewable. Inspect countershaft closely for any scoring or wear of bearing surfaces or damage to teeth of final drive gear machined on rear end of shaft. Renew countershaft if any of these defects are noted. If renewing bushing (62), ream if necessary to provide 0.003-0.005 clearance between bushing and sleeve (64) on pto shaft.

Reassemble countershaft by reversing disassembly procedure. Note that front connector (43) has recess for thrust washer (42), and must be installed with recess towards front end of shaft. Install rear connector (53) with small identifying notches in splines towards rear end of shaft. Sliding couplings (44, 48 and 54) are interchangeable (not matched with connectors) and can be installed in either direction. Refer to Fig. FO153 for guide to assembling gears on countershaft.

206. Bushing (33—Fig. FO153) and bearing cup (35) in front bearing retainer (32) are renewable. Remove bearing cup with bearing cup puller and remove bushing with chisel. Ream new bushing after installation to provide 0.003-0.005 clearance between bushing and hub of PTO driven gear (37). Cone and roller assembly (36) can be removed from PTO driven gear with long, thin punch inserted through access holes in hub of gear; bearing cup (38) can be removed from gear with bearing cup puller attachment on slide hammer.

Refer to paragraph 290 for servicing PTO support (69) and shifter assembly which includes bearing cup (61) and oil seal (65).

207. Reinstall countershaft and gear assembly by reversing removal procedures outlined in paragraph 204, but do not install PTO shaft (63—Fig. FO153) at this time. Install PTO support (69) with shim stack (68) and tighten retaining screws to a torque of 25-30 Ft.-Lbs. Rap the support sharply a few times with a soft-faced hammer to seat the bearings. Wrap a cord around the final drive gear teeth on rear end of countershaft (59) and attach a pull scale to the cord. Bearing adjustment (preload) is correct if a steady pull of 6-8 pounds is required to rotate countershaft when unit has been assembled with all bearing cups, cone and roller assemblies re-used, or a steady pull of 14-16 pounds if all new bearing cups and cone and roller assemblies have been installed. Add to shim stack (68) if pull required to rotate shaft is greater than specified, or remove shims from stack if pull is less than specified. Shims are available in thicknesses of 0.003, 0.005 and 0.012.

After correct bearing adjustment has been established, remove PTO support and install PTO shaft (63). Reinstall support with correct shim stack thickness and tighten retaining cap screws to a torque of 25-30 Ft.-Lbs.

79

"SELECT-O-SPEED" TRANSMISSION

The "Select-O-Speed" transmission is a planetary gear drive unit providing ten forward and two reverse speeds. Desired gear ratio is selected by moving a control lever and starting, stopping or changing gear ratios is accomplished without operation of a conventional clutch. A foot operated feathering valve is provided for interrupting the gear train in case of emergency or for close maneuvering such as the hitching or unhitching of implements.

The transmission is available in three types, all basically similar except for PTO options. The transmission is available without a power take-off, with a 540 RPM engine drive power take-off, or with 540-1000 RPM engine drive and a ground speed ratio power take-off.

For a better understanding of the operation of the "Select-O-Speed" transmission, fundamental operating principles of a planetary gear system are outlined in following paragraphs 210 to 217.

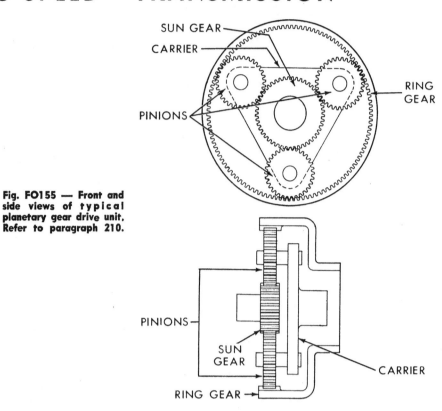

Fig. FO155 — Front and side views of typical planetary gear drive unit. Refer to paragraph 210.

PLANETARY GEAR POWER FLOW

210. Refer to Fig. FO155. The three "elements" of a planetary system are the sun gear, pinion carrier and the ring gear. When any element is rotated, the other two elements will also turn unless one is held by an external force. Depending upon which element is held, power can be applied or taken out at the sun gear, pinion carrier or ring gear. The possible means of obtaining different gear ratios from a planetary gear set are as follows:

211. Turning the sun gear and holding the ring gear forces the pinions to turn within the ring gear moving the pinion carrier with them. Thus, the carrier turns in the same direction as the sun gear but at a slower speed. (Underdrive ratio)

212. Turning the ring gear and holding the sun gear forces the pinions to turn within the ring gear moving the pinion carrier with them. The carrier turns in the same direction as the ring gear, but at a slower speed. (Underdrive ratio)

213. Turning the pinion carrier and holding the ring gear forces the pinions to turn within the ring gear causing the sun gear to turn in the same direction as the carrier, but at a higher speed. (Overdrive ratio)

214. Turning the pinion carrier and holding the sun gear forces the pinions to turn around the sun gear causing the ring gear to turn in the

same direction as the carrier, but at a higher speed. (Overdrive ratio)

215. Turning the sun gear and holding the pinion carrier forces the pinions to act as idlers turning the ring gear in opposite direction from sun gear and at a slower speed. (Underdrive reverse ratio)

216. Turning the ring gear and holding the pinion carrier forces the pinions to act as idlers turning the sun gear in the opposite direction from ring gear and at a higher speed. (Overdrive reverse ratio)

217. Locking any two units of a planetary system together results in a solid drive unit, and if any element is turned, all three elements turn in the same direction at the same speed. (Direct drive ratio)

TRANSMISSION OPERATION

218. **PLANETARY SYSTEMS.** The "Select-O-Speed" transmission utilizes four planetary systems designated, from front to rear, as "A", "B", "C" and "D". Elements of each planetary system are designated as "A" sun gear, "A" carrier, "A" ring gear, etc. Refer to Fig. FO156.

The "D" (rear) planetary system is used as a final reduction unit only. Power always enters at the "D" sun

gear, the "D" ring gear is held stationary in the transmission case and power is taken out from the "D" carrier which is integral with the transmission output shaft.

Six basic speed ratios, five forward and one reverse, are obtained by various combinations of holding or applying power to different elements of the "B" and "C" planetary systems.

The "A" (front) planetary system is used as a direct drive-overdrive unit to double the basic speed ratios providing ten forward and two reverse speeds.

219. **PLANETARY CONTROLLING UNITS.** Three brake bands, three multiple disc clutches and either a one way (overrunning) clutch (1959-62 production transmissions) or a fourth multiple disc clutch (1963 and later production transmissions) are used to control the "A", "B" and "C" planetary systems and provide the different speed ratios.

The three brake bands are designated, from front to rear, as Band 1, Band 2 and Band 3. The bands are actuated by hydraulic servos designated Servo 1, Servo 2 and Servo 3 to correspond with the three bands. The servos contain springs that work in the opposite direction from hydraulic pressure. Servo 1 applies Band 1 with

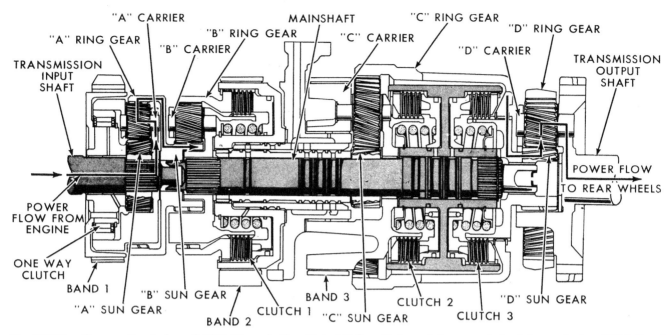

Fig. FO156 — Cross-sectional view of 1959-62 production "Select-O-Speed" transmission gear train. Late production (1963 & up) transmission is similar except that a multiple disc clutch is used instead of the one-way clutch shown at front end of power train.

hydraulic pressure and releases the band with spring pressure. Servos 2 and 3 apply Bands 2 and 3 with spring pressure and release the bands with hydraulic pressure. The bands, when applied, hold planetary elements stationary as follows:

Band Applied	Planetary Element Held
1	"A" Sun Gear
2	"B" Ring Gear and "C" Sun Gear
3	"C" Carrier

In early (1959-62) transmissions, the one way (overrunning) clutch automatically locks the "A" carrier to the "A" sun gear whenever Band 1 is released. On later production (1963 and up) transmissions, a multiple disc clutch is used instead of the one way clutch.

The multiple disc clutch packs are designated, from front to rear, as Direct Drive Clutch (1963 and later, only), Clutch 1, Clutch 2 and Clutch 3. The clutches are engaged by hydraulic pressure and disengaged by spring pressure. The clutches, when engaged, lock planetary elements together as follows:

Clutch Engaged	Planetary Elements Locked
Direct Drive*	"A" Sun Gear to "A" Carrier
1	"B" Carrier to "B" Ring Gear
2	"B" Carrier to "C" Carrier
3	"B" Carrier to "D" Sun Gear

*Direct Drive clutch used on 1963 and later transmissions only.

"SELECT-O-SPEED" HYDRAULIC SYSTEM

The hydraulic system required to actuate the planetary controlling bands and clutches and the power take-off clutch is contained within the transmission, and transmission lubricating oil is utilized as hydraulic fluid for the system. A roller vane pump, located on the front cover and driven by the transmission input shaft, supplies system pressure. A multiple valve control unit within the transmission directs pressure to the servos and clutch pistons as required to provide the different speed ratios, to start and stop the tractor, to operate the power take-off clutch and also regulates circuit pressures within the system.

220. **HYDRAULIC CIRCUITS, 1959-62 PRODUCTION UNITS.** Refer to Fig. FO157 for diagram of the hydraulic circuits within the control valve used in 1959-62 production transmissions.

Pressure is supplied directly to the Band 2 and Band 3 control valves (direct transmission circuit), indirectly to the Clutch 1, Clutch 2, Clutch 3 and Band 1 control valves via the transmission regulating valve and feathering valve (indirect transmission circuit), indirectly to the PTO feathering valve via the PTO regulating valve (indirect PTO circuit) and indirectly to the lubrication valve via the system relief valve (indirect lubrication circuit).

Refer to Fig. FO159 for diagrams showing operation of the transmission and PTO feathering valves. With the inching pedal up or the PTO control handle pulled out, pressure from the regulating valves is directed to the indirect transmission circuit or to the PTO clutch. With the inching pedal fully depressed or the PTO control handle pushed in, the feathering valves block pressure from the indirect transmission circuit or PTO clutch and opens the indirect transmission circuit or PTO clutch pressure line to the sump. With the feathering valves in the mid-position (inching pedal partially depressed or PTO control handle partly pulled out), pressure in the indirect transmission circuit or PTO clutch can be "feathered" to provide smooth starting of tractor or engagement of PTO clutch. This is similar to slipping a conventional clutch.

The transmission regulating valve and the PTO regulating valve prevent oil pressure in the direct hydraulic circuit from dropping below the valve pressure setting (145-155 psi) whenever either or both feathering valves are in the mid or "feathering" position. The system relief valve limits pressure to 170-190 psi within the system. The lubrication valve maintains, at the valve, about 43 psi in the transmission lubrication circuit. (Lubrication circuit pressure at the pressure switch location on rear cover

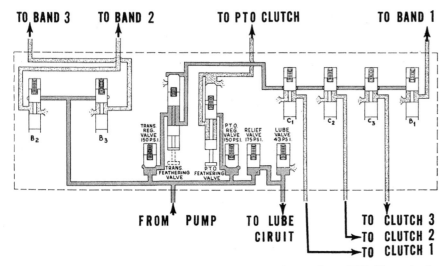

Fig. FO157 — Hydraulic flow diagram of control valve used on 1959-62 production transmission. A later type control valve (See Fig. FO158) can be installed in earlier transmissions.

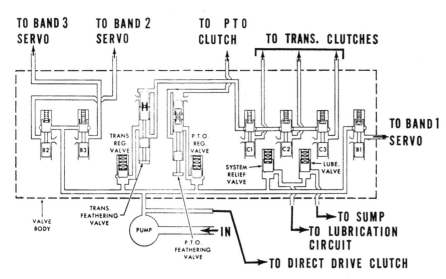

Fig. FO158 — Hydraulic flow diagram of control valve used on late (1963 & up) production transmissions.

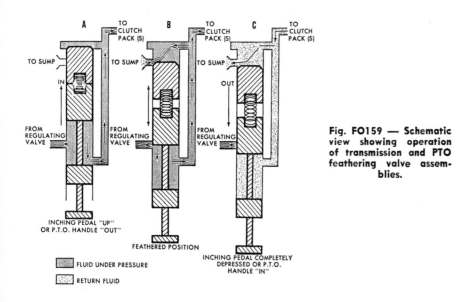

Fig. FO159 — Schematic view showing operation of transmission and PTO feathering valve assemblies.

should be 2-12 psi due to normal "leakage" in the circuit.) As pressure in the system drops to 145-155 psi when either feathering valve is in mid-position, no oil will pass the 170-190 psi system relief valve; thus, lubrication circuit pressure drops to zero at that time. Oil by-passing the lubrication valve returns to the sump through the oil filter.

To prevent Band 2 and Band 3 from being applied at the same time, and locking up the transmission when changing speed ratios, an interlock circuit between Servo 2 and Servo 3 is required. Refer to Fig. FO160 for diagram of the interlock circuit. Band 2 and Band 3 are applied by servo spring pressure and released by hydraulic pressure on the servo piston. However, before Band 3 can be applied, Band 2 must be released which opens the interlock check valve and allows oil trapped in Servo 3 to return to the sump. The opposite is also true; to apply Band 2, Band 3 must be released moving the valve attached to Servo 3 so that oil trapped in Servo 2 can return to the sump. The interlock valves are located in the cover on left side of transmission case.

NOTE: The control valve described in paragraph 221 may be used to service 1959-62 production transmissions. However, the 1959-62 production valve cannot be used to service 1963 and later transmissions.

221. HYDRAULIC CIRCUITS, 1963 AND LATER PRODUCTION UNITS. Refer to Fig. FO158 for diagram of hydraulic circuits within the control valve used on 1963 and later production transmissions. Other than Band 1 control valve being included in the direct transmission circuit, the control valve assembly is similar to that described in paragraph 220 for 1958-62 production transmissions.

The interlock circuit described in paragraph 220 is also used in 1963 and later production transmissions.

In addition, a shuttle valve operate by pressure in the Servo 1 pressure line is used in 1963 and later production transmissions to actuate the Direct Drive Clutch. Refer to Fig. FO161 for diagram of the shuttle valve circuit. When Band 1 is applied, pressure in the Servo 1 line moves the shuttle valve to block pressure to the Direct Drive Clutch and allow oil trapped in the clutch to return to the sump. Thus, whenever Band 1 is applied, the Direct Drive Clutch is released. When Band 1 is released,

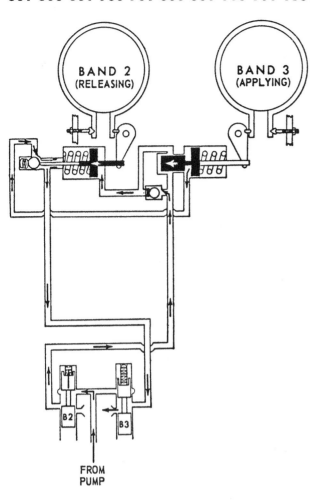

Fig. FO160 — Schematic diagram showing hydraulic interlock circuit between Band 2 and Band 3. Oil cannot return from Servo 3 and allow spring pressure to apply Band 3 until Band 2 is released and Servo 2 opens the check valve to allow oil to return to sump. The opposite is also true; Band 2 cannot be applied until Band 3 is released and the plunger valve on Servo 3 moves back to allow oil trapped in Servo 2 to return to the sump.

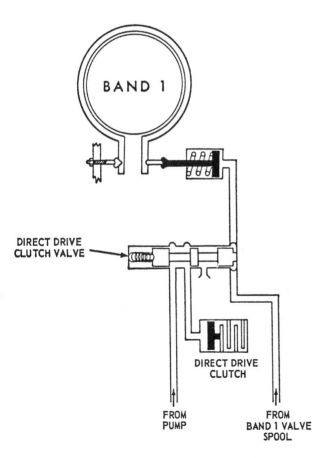

Fig. FO161 — On late production (1963 & up) transmissions, a shuttle valve (Direct Drive Clutch Valve) allows oil pressure directly from the transmission pump (See Fig. FO158) to engage the Direct Drive Clutch whenever Band 1 is released. When oil pressure from Band 1 valve spool starts to build up in the line to Servo 1, the shuttle valve blocks off the oil pressure from pump and allows oil in the Direct Drive Clutch piston to return to the sump, disengaging the Direct Drive Clutch just before Band 1 is applied.

pressure in the Servo 1 line drops to zero and the spring behind the shuttle valve moves it into position shown in Fig. FO161 allowing oil pressure to engage the Direct Drive Clutch.

222. **CONTROL POSITIONS.** Fourteen control positions are provided on the "Select-O-Speed" gear selector. Units of the transmission activated in each control position are indicated on the Trouble Shooting Charts in Fig. FO172 for 1958-62 production transmissions and Fig. FO173 for 1963 and later production transmissions.

Shaded areas in Band and Clutch unit columns of Trouble Shooting Charts indicate which control positions of control valve direct hydraulic pressure to the units.

Bold face "A" in Band and Clutch unit columns indicates selector positions in which the unit is applied or engaged.

Malfunction conditions are also indicated in the Trouble Shooting Charts.

SYSTEM ADJUSTMENTS

Malfunctions of the "Select-O-Speed" Transmission, for which service is required, could be from a number of causes; the most common of which, will be maladjustment of one or more units of the transmission correctable by a complete unit adjustment. The first step in correcting troubles, therefore, would be a complete operational adjustment on the three transmission bands, the four pressure regulating valves and the selector assembly. As part of the adjustments are made with the engine running and the selector lever in an operational position, the first step is to disengage the traction coupling as follows:

223. **TRACTION COUPLING.** All tractors equipped with a "Select-O-Speed" transmission incorporate a traction coupling sleeve which can be shifted to disengage the transmission output shaft from the differential pinion. The shift-lever shaft is in the inspection plate located on the left side of the rear axle center housing (Fig. FO162). To disengage, unbolt and remove the shaft retainer, and using retainer as a handle, turn shaft clockwise to disengage coupling. Reinstall retainer to hold coupling in disengaged position. Traction coupling must also be disengaged if it is necessary to tow tractor.

224. **FLUID LEVEL CHECK.** Before an attempt is made to start or service the tractor, first check the fluid level.

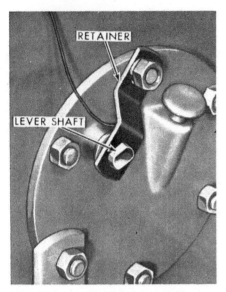

Fig. FO162 — Left side of rear axle center housing showing traction coupling disengaging mechanism. Traction coupling must be disengaged for system adjustments or for towing tractor.

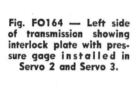

Fig. FO163 — Right side of transmission showing Band 3 adjusting screw, Servo 1 cover and transmission fluid level and filler plug.

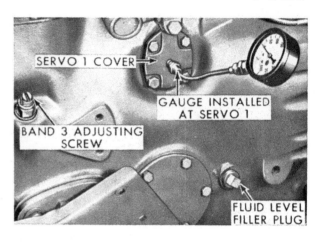

Fig. FO164 — Left side of transmission showing interlock plate with pressure gage installed in Servo 2 and Servo 3.

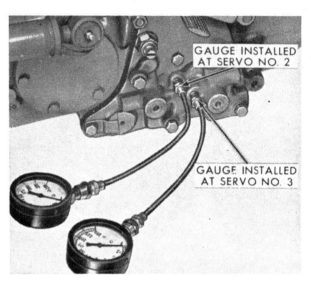

To check, remove the pipe plug located on the right side of the transmission housing (Fig. FO163). Fluid should be at the plug with tractor standing level. If fluid is low, fill to plug level, through plug opening, with Ford Hydraulic Fluid, Specification M-2C-41

225. PRESSURE CHECK. As stated in paragraphs 220 and 221 and illustrated in Figs. FO157 and FO158, the transmission hydraulic system is divided into four separate but interrelated circuits as follows:

a. Direct Transmission Circuit serving Bands 2 and 3 (and Band 1 on 1963 and later production units), and the other three systems through their respective regulating valve.

b. Indirect Transmission Circuit, serving the three clutches (and Band 1 on 1959-62 production units) and connected to the direct transmission circuit by the 145-155 psi transmission regulating valve.

c. Indirect PTO Circuit (on Models with PTO), serving the pto clutch and connected to the direct transmission circuit by the 145-155 psi power take-off regulating valve.

d. Transmission Lubrication System, with lubrication passages to the three multiple disc clutches, the one-way clutch and the shaft bearings. The lubrication circuit is connected to the direct transmission circuit by the 170-190 psi system relief valve, and protected by its own 43 psi lubrication pressure valve.

Provisions are made for installation of a pressure gage in servos 1, 2 and 3 as illustrated in Fig. FO163 and Fig. FO164. Gages may be used in all three locations; however, all pressure valve adjustments and a complete systems diagnosis can normally be made by the installation of a gage in No. 2 servo. Following procedure is for pressure check using one gage; however, same conclusions can be made with gages in all three locations.

NOTE: Tractors produced in late 1962 and early 1963 do not have a "Select-O-Speed" lubrication indicator lamp. To make pressure check on these tractors, it will be necessary to install a zero to 25 psi gage in the lubrication circuit as follows: Remove traction coupling shifter plate from left side of center housing (or, on tractors without hydraulic lift system, remove center housing top cover plate). Remove pipe plug from upper left side of transmission output shaft opening on transmission rear support plate. Using a flexible hose and elbow fitting, install gage in pipe plug opening. Gage should read 2-12 psi when inching pedal is

released or fully depressed, or when PTO control is fully in or out, and drop to zero when either the inching pedal or PTO control is in the "feathering" or mid position. This will correspond to normal action of the lubrication indicator lamp.

To check the pressure, first disengage the traction coupling as outlined in paragraph 223, bring the transmission fluid up to operating temperature and install a 0-300 psi gage in Servo 2 location. Shift selector lever to neutral position and operate engine at 800 RPM. The pressure gage should read 170-190 psi. A higher reading would indicate that the system relief valve setting is incorrect and valve will need to be adjusted as outlined in paragraph 226. If gage reading is below 170 psi, stop engine and move gage to Servo 3 location. Start engine, move speed selector to 5th speed position and compare gage reading with that made at Servo 2 location. If Servo 3 reading is higher, it would indicate leakage at Servo 2, and the Servo 3 loca-

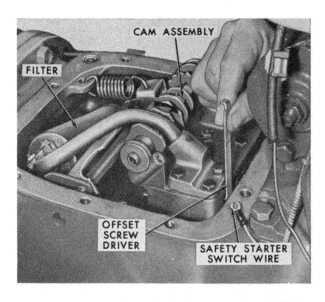

Fig. FO165 — View showing method of adjusting the regulating valves.

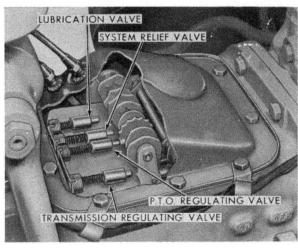

Fig. FO166 — Cut-away view showing location of regulating valves.

JUSTMENT. To adjust the regulating valves, first remove the transmission cover as outlined in paragraph 234 and the control valve as outlined in paragraph 238; then, proceed as follows:

Remove the two adjusting screw retainers from front of upper valve body; remove the adjusting screws and withdraw the regulating springs and valves. Polish valves and bores, if necessary, to make sure valves operate smoothly, clean the valves and valve bores; then reinstall valves, valve springs and adjusting screws.

NOTE: The four valves are interchangeable. The lubrication valve (light) spring is to the right when valve is installed. (See Fig. FO166). The other three springs are interchangeable. On models without power take-off, a pin is used instead of spring in PTO regulating valve to lock valve in closed position.

An approximately correct bench adjustment of the adjusting screws can be made as follows: Starting from end of valve body opposite the cable wheel, tighten the first (transmission regulating valve) adjusting screw flush with valve body; tighten second (PTO regulating valve) adjusting screw flush with body; the third (system relief valve) adjusting screw flush plus one (1) turn; and the fourth (lubrication valve) screw all the way in until it bottoms. Note: On models without PTO, turn second (PTO regulating valve) screw in tight against pin.

Reinstall the control valve as outlined in paragraph 238, but without the retainers; then recheck and adjust pressures as follows: Ground out the safety starter switch and restart the tractor. Rotate the control cam assembly forward three positions to neutral (N) and adjust the regulating valves by turning the adjusting screws in or out as required. Fig. FO166 illustrates the location of the four regulating valves. The long adjusting screw retainer makes a good screwdriver for adjusting the valves.

If it is impossible by means of system regulating valve adjustment, to obtain the 170-190 psi system operating pressure, first check the lubricatin indicator lamp with the engine running. If the light is off, stop the tractor and turn the ignition switch back to the "On" position. If the light then comes on it would indicate that the lubrication system is functioning and the trouble lies in the system relief valve, and the valve must be removed and serviced as outlined in

tion and 5th speed position should be used for remainder of pressure check. If gage reading at Servo 3 location is still below 170 psi, readjust system relief valve as outlined in paragraph 226 before proceding with pressure check.

If gage reading is 170-190 psi at Servo 2 or Servo 3 location, partially depress inching pedal until lubrication lamp flashes on, or gage at rear support drops to zero. Pressure gage reading at Servo 2 or 3 should then be 145-155 psi. A reading higher than 155 psi or lower than 145 psi will indicate that transmission regulating valve setting is incorrect and will need to be reset.

Release inching pedal and gradually pull PTO control handle out until lubrication lamp flashes on or gage at rear support drops to zero. Gage reading at Servo 2 or 3 should then be 145-155 psi. A reading higher than 155 psi or lower than 145 psi will indicate that the PTO regulating valve

setting is incorrect and will need to be reset.

Lubrication relief valve setting is correct if pressure gage at rear support reads 2-12 psi when gage reading at Servo 2 or Servo 3 is above 170 psi. If tractor is equipped with lubrication indicator lamp, lubrication relief valve setting can be assumed to be correct if lamp reacted normally in preceding checks. That is, lamp should flash on when gage reading at Servo 2 or 3 drops to 145-155 psi, and go out when the gage reading returns to above 170 psi. If lamp remains on at all times, it would indicate a faulty pressure switch (located on transmission rear support plate), grounded wire, inoperative lubrication relief valve, or excessive leaking in lubrication circuit. If indicator lamp should fail to light up, it would indicate a burned out bulb, loose connection, blown fuse, broken wire or faulty pressure switch.

226. REGULATING VALVE AD-

paragraphs 238 and 239. If the lubrication indicator lamp remains on and the 170-190 psi pressure cannot be obtained, it would indicate that the pump is at fault or a leak exists in the system. Start the engine and examine the transmission for evidence of leaks or oil turbulence which would indicate a leak. If none are found, the pump will need to be removed and serviced as outlined in paragraph 241. If leaks are found, the transmission must be disassembled and the leaks repaired before proceeding further.

When proper adjustment has been made on the regulating valves, remove the control valve and reinstall the adjusting screw retainers. If adjustment of the regulating screw slots is necessary for retainer installation, turn the adjusting screws inward to increase the pressure until proper alignment is obtained. Reinstall the control valve, filter manifold, starter switch and the transmission cover.

227. BAND ADJUSTMENTS. Bring transmission to operating temperature, disengage the traction coupling as outlined in paragraph 223 and proceed as follows:

Band No. 1: With engine **not** running, loosen the adjusting screw lock nut at least two full turns while holding adjusting screw stationary. With torque wrench and screwdriver (drag link) socket, tighten the adjusting screw to 10 Ft.-Lbs. Check to make sure that lock nut did not seat on transmission housing (sealing washer still loose); if washer is tight, back-off lock nut farther and re-tighen screw with torque wrench. Then, back adjusting screw out exactly one full turn. Start engine, move speed selector to 3rd speed position, hold adjusting screw stationary and tighten lock nut to a torque of 20-25 Ft.-Lbs.

Bands 2 and 3 are adjusted with engine running at 800 RPM as outlined in following paragraphs:

Band No. 2: Shift transmission to Park with engine running; then, hold adjusting screw stationary while backing off locknut two full turns. Shift transmission to Neutral and with a torque wrench tighten adjusting screw to a torque of 5-10 Ft.Lbs. while observing the precautions outlined for band No. 1. Back off adjusting screw exactly ¾-turn; shift transmission to Park position, hold screw stationary and tighten locknut. Band 2 adjusting screw is located on the interlock plate on the left side of the transmission case as illustrated in Fig. FO169.

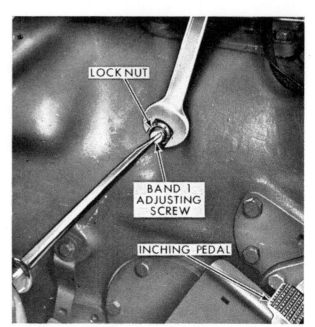

Fig. FO167 — Band No. 1 adjustment. Loosening locknut.

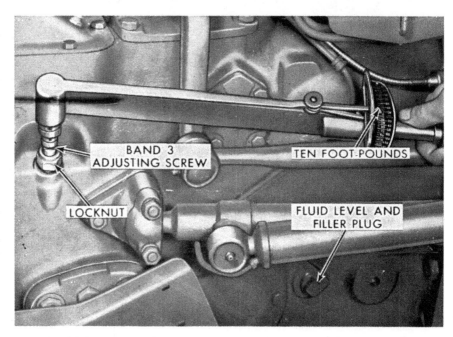

Fig. FO168 — Band No. 3 adjustment. Adjusting Band 3 with torque wrench.

Band No. 3: Shift selector lever to Park, hold adjusting screw and back off locknut at least two full turns. Shift transmission to fifth speed and tighten the adjusting screw to a torque of 5-10 Ft.-Lbs. as illustrated in Fig. FO168. Back off the screw exactly ¾-turn; shift transmission to Park position, and tighten locknut.

NOTE: If engine stalls before a torque of 5 Ft.-Lbs. is obtained, stop tightening just before engine stalls; then back off ¾-turn.

228. SELECTOR ADJUSTMENT. For positive identification of speed selections, the individual speed indications on the dial should always be positioned directly under the pointers in the housing. Adjustment for wear or misalignment can be made as follows:

Remove the left-hand selector shaft cover as illustrated in Fig. FO170, move the selector lever to the neutral position and loosen the shaft nut

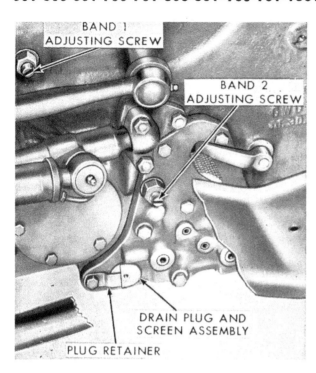

BAND 1 ADJUSTING SCREW

BAND 2 ADJUSTING SCREW

DRAIN PLUG AND SCREEN ASSEMBLY

PLUG RETAINER

Fig. FO169 — Left side of transmission showing location of Band 1 and Band 2 adjusting screws. Drain plug location shown is for 1959-62 production transmissions only. Refer to paragraph 232.

pedal is released one of five conditions will prevail.

(1) The transmission will operate properly in that control position.

(2) The transmission will operate in an incorrect speed ratio.

(3) The transmission will go to neutral in control position other than neutral.

(4) The transmission will go to park in control position other than park.

(5) The transmission will lock up (stall engine).

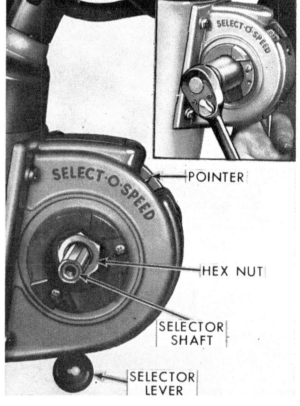

POINTER

HEX NUT

SELECTOR SHAFT

SELECTOR LEVER

Fig. FO170 — Selector assembly with left cover removed showing method of aligning selector dial. Note: If hex nut does not loosen easily, tap wrench sharply with hammer to provide impact. Excessive torque will twist the die-cast selector shaft.

while holding the lever firmly in the detent as shown in the inset. Move the dial to proper alignment and re-tighten nut.

Positive lever stops may be installed on 3rd, 5th and 7th speeds and the two reverse speeds in order to facilitate the location of these speeds. Stop screws not in use are installed in the selector side plate opposite the lever. To install, place the screws in the desired tapped holes as illustrated in Fig. FO171. The control lever may be moved to the left side of the selector dial if desired.

TROUBLE-SHOOTING

229. OPERATIONAL CHECK. If the system adjustments outlined in the previous section fail to correct transmission malfunctions, the next step in trouble diagnosis would be an operational check. To perform this check, the traction coupling must first be engaged by reversing the steps outlined in paragraph 223, then proceed as follows:

Start the tractor and set the engine speed at 800 rpm. Put the transmission into neutral by depressing the inching pedal, shift the selector lever into each of the 14 control positions in turn and gradually release the inching pedal. Note the reaction in each position for later reference to the Trouble-Shooting Chart in Fig. FO172 or Fig. FO173. When the inching

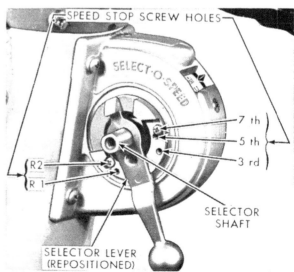

SPEED STOP SCREW HOLES

7 th

5 th

3 rd

R2

R1

SELECTOR SHAFT

SELECTOR LEVER (REPOSITIONED)

Fig. FO171 — Left side of selector assembly with lever repositioned. Speed stop screw holes are located on both sides of selector for installation of speed stop screws.

If transmission operates properly in all control positions, proceed with PRESSURE CHECKS as outlined in paragraph 231.

If any of the other conditions, (2), (3), (4) or (5) are encountered, compare the notes made during operating check with the inoperative conditions listed in each band and clutch unit column of the proper Trouble-Shooting Chart in Fig. FO172 or Fig. FO173. If transmission is a 1959-62 production unit, refer to Fig. FO172; refer to Fig. FO173 for 1963 and later production units.

As an example, assume the notes made during the operational check of a 1963 production transmission are as follows:

Control Position	Condition Encountered
P	N
R2	R2
R1	R1
N	N
1	1
2	2
3	3
4	4
5	N
6	N
7	N
8	N
9	9
10	10

NOTE: P — indicates Park, N — indicates Neutral, L — indicates lock-up, R — indicates reverse and numbers indicate gear ratios.

In comparing the conditions assumed for the example with the Trouble-Shooting Chart in Fig. FO173 for 1963 and later production transmissions, it is found that the conditions compare to the inoperative conditions in the Band 2 unit column indicating cause of trouble is Band 2 remaining released in all selector positions.

230. TORQUE LIMITING CLUTCH.
A torque limiting clutch is installed in the engine flywheel and functions as an overload clutch. Slippage of the torque limiting clutch under normal loads will cause an interruption or lowering of transmission pump flow and system pressure and a lockup may occur. A defective torque limiting clutch is to be suspected if the transmission malfunction exists only under extremely heavy loads, espe-

POSITION OR GEAR	OVER RUNNING CLUTCH (a)	*** B1 * (r a)	DIRECT CIRCUIT BAND B2 ** (r a)	DIRECT CIRCUIT BAND B3 ** (r a)	INDIRECT CIRCUIT CLUTCH PACK C1 * (r a)	INDIRECT CIRCUIT CLUTCH PACK C2 * (r a)	INDIRECT CIRCUIT CLUTCH PACK C3 * (r a)
P	P A P	P P	A N	P A N	L P	L P	L P
R2	L R2	R2 A R1	L R2	R2 A N	R2 A N	L R2	R2
R1	R1 A N	R2 R1	L R1	R1 A N	R1 A N	L R1	R1
N	N A N	N N	P N	N A N	R1 N	2 N	1 N
1	1 A N	3 1	L 1	1 A N	L 1	L 1	1 A N
2	2 A N	4 2	L 2	2 A N	L 2	2 A N	L 2
3	L 3	3 A 1	L 3	3 A N	L 3	L 3	3 A N
4	L 4	4 A 2	L 4	4 A N	L 4	4 A N	L 4
5	5 A N	7 5	5 A N	L 5	L 5	5 A N	5
6	6 A N	8 6	6 A N	L 6	6 A N	L 6	6
7	L 7	7 A 5	7 A N	L 7	L 7	7 A N	7
8	L 8	8 A 6	8 A N	L 8	8 A N	L 8	8
9	9 A N	10 9	9 A N	L 9	9 A N	9 A N	9
10	L 10	10 A 9	L 10	L 10	10 A N	10 A N	10 10

A—Indicates unit applied for that selector position.
L—Indicates lock-up condition.
a—Indicates unit remains applied.
r—Indicates unit remains released.
*—Indicates spring applied unit.
**—Indicates pressure applied unit.
***—Note: Band 1 can be in either indirect or direct transmission circuit. Refer to text.

Fig. FO172 — Trouble-shooting chart for 1959-62 production "Select-O-Speed" transmissions. Shaded areas in Band or Clutch columns indicates selector position in which hydraulic pressure is directed to that unit.

cially in the higher speed ranges and the transmission operates properly when shifted to a lower speed range.

To check the torque limiting clutch start the engine and bring the tractor up to operating temperature. Shift the selector lever into 8th, 9th or 10th speed position, increase the engine speed to full throttle and quickly apply both brakes. If the tractor forward motion can be halted without pulling the engine down below 1000 RPM or stalling the engine, the torque limiting clutch will need to be renewed as outlined in paragraph 240.

231. PRESSURE CHECKS.
Leakage in any of the clutches or servos will only occur when that unit is actuated. To check the various units for leakage, first disengage the traction coupling as outlined in paragraph 223, insert 0-300 psi pressure gages in Servos 1, 2 and 3. (Note: Diagnosis for pressure leaks can be made by using only one gage, but will require moving the gage to the different Servos.) With transmission at operating temperature and engine running at 800 RPM, proceed as follows:

Move the speed selector through each of the 14 positions and make a record of the Servo gage readings in each position. All readings should be within 170-190 psi, with a differential of 3 psi or less between readings, in accordance with the following chart:

PRESSURE GAGE READINGS

Gear Ratio	Band 1 Servo	Band 2 Servo	Band 3 Servo
Park	0	0	0
R2	180	180	180
R1	0	180	0
N	0	180	0
1	0	180	0
2	0	180	0
3	180	180	0
4	180	180	0
5	0	0	180
6	0	0	180
7	180	0	180
8	180	0	180
9	0	180	180
10	180	180	180

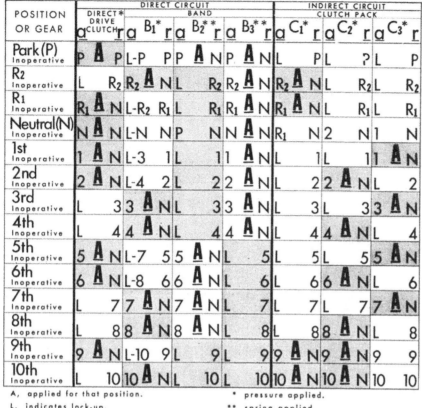

A, applied for that position.
L, indicates lock-up.
a, applied and does not release.
r, released and does not apply.

* pressure applied.
** spring applied.
☐ hydraulic pressure on.

Fig. FO173 — Trouble-shooting chart for 1963 and later production transmissions. Note that two different inoperative conditions are given for some selector positions in the Band 1 applied column. First condition (lock-up) occurs if Band 1 remains applied due to mechanical malfunction. Second condition occurs if Band 1 remains applied due to hydraulic malfunction.

If pressure readings are consistently higher than 190 psi, or lower than 170 psi, readjust the system relief valve as outlined in paragraph 226. If readings are low and cannot be increased by adjusting system relief valve, and no leakage is evident, overhaul pump as outlined in paragraph 241.

If pressure readings are low only in certain gear ratios, move the selector lever to each position having low readings and depress the inching pedal to step plate. If pressure readings increase to that observed in selector position having high pressure gage readings, leakage is occurring in one of the units in the indirect transmission circuit. If pressure readings do not increase when inching pedal is depressed, leakage is occurring in the direct transmission circuit. As a unit will leak only when pressure is directed to that unit, compare the selector positions having low readings with the shaded (pressure on) markings in unit columns of the Trouble Shooting Chart in Fig. FO172 or Fig. FO173.

As an example, assume the pressure gage readings indicated leakage in the indirect transmission circuit and the low pressure gage readings occurred in selector positions 2, 4, 6, 8, 9 and 10. Comparing this information with the Trouble Shooting Chart will show that Clutch 2 is the unit involved since the Clutch 2 column of the chart is the only one with shaded marking in the selector positions matching the pattern of low pressure readings.

NOTE: Band 1 is in the direct transmission circuit on 1963 and later production transmissions. However, Band 1 can be in either the direct transmission circuit or indirect transmission circuit on 1959-62 transmissions depending on whether transmission is equipped with late type or original (1959-62) type control valve. If, when inching pedal is depressed, gage reading at Servo 1 drops from 170-190 psi to zero, transmission is equipped with original type valve with Band 1 in the indirect transmission circuit.

If gage reading remains at 170-190 psi with inching pedal depressed, transmission is equipped with late type control valve with Band 1 in the direct transmission circuit.

To check for leakage in the power take-off circuit, move selector lever to any position giving 170-190 psi reading on any of the Servo 1, 2 or 3 gages and pull PTO control handle out to the engaged position. If gage reading decreases, leakage is occurring in the power take-off circuit or clutch.

TRANSMISSION OVERHAUL

Access to the control valve assembly is obtained by removal of the transmission top cover. Access to the hydraulic pump, torque limiting clutch and front adapter plate (cover) is obtained by splitting the tractor between the transmission housing and engine assembly. To service the input shaft or pto system, the transmission must be removed from the tractor. All other transmission components may be removed and reinstalled from the rear of the transmission case after detaching the transmission housing from the rear axle center housing. The first step in transmission overhaul is to determine, as nearly as possible, which components require service and remove only the parts necessary to obtain access to those components.

Note: The "Select-O-Speed" transmission is a hydraulic unit and merits the same statndards of care and cleanliness accorded any hydraulic or diesel unit. Disassembly or service should only be attempted in a clean, dust free shop and the removed assemblies stored and serviced only where good housekeeping is observed.

CAUTION: Never use cloth shop towels or rags to wipe internal parts of the "Select-O-Speed" transmission. Lint from cloths or rags will bind the closely fitted control valve parts, clog oil passages and possibly cause serious damage and/or improper transmission operation.

232. **DRAINING THE TRANSMISSION.** To drain the transmission on 1959-62 production tractors, loosen the cap screw holding the drain plug retainer on the right side of the transmission case (see Fig. FO169). Insert a 1/4-20 bolt into drain plug threads and withdraw plug sufficiently to expose the "O" ring. Rotate the plug 180 degrees and withdraw the plug until the drain hole is exposed. Before drain plug is reinstalled, remove plug completely and clean the strainer screen.

On 1963 and later production, trans-

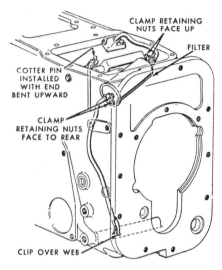

Fig. FO174 — View showing transmission oil filter installation.

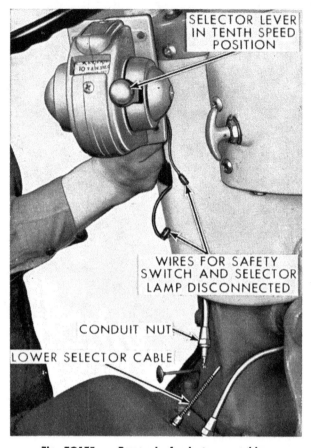

Fig. FO175 — Removal of selector assembly.

mission is provided with a ½-inch pipe thread drain plug in bottom of case. After draining the transmission, remove and clean the strainer screen following procedure outlined in preceding paragraph.

233. **TRANSMISSION OIL FILTER.** The transmission oil filter can be renewed after removing the transmission top cover as outlined in paragraph 234. To remove filter, unbolt filter tube manifold from control valve housing and withdraw filter, manifold tube and outlet tube from transmission.

Loosen clamp bolts and remove old filter can from manifold and outlet tubes. When installing new filter, refer to Fig. FO174 and proceed as follows: Be sure that end of filter can marked "IN" is attached to manifold tube and attach outlet tube to other end of filter can. Insert outlet tube down inside left rear corner of transmission case and be sure clip on tube is hooked over web in bottom of case. Filter clamps must be turned as shown in Fig. FO174 to avoid interference. If inching pedal rod contacts outlet tube, move lower end of tube towards center of transmission by prying against bottom end of tube with long screwdriver. Using new gasket, bolt filter manifold tube to control valve. Make final check to be sure there is no interference and reinstall transmission top cover.

234. **R&R TOP COVER.** To remove the transmission top cover, it is first necessary to remove the selector assembly as shown in Fig. FO175, as follows:

Place transmission in park position. Completely loosen the conduit nut securing the selector cable conduit to the transmission cover. Remove the screws securing the selector assembly to the tractor hood. Move the selector lever upward to tenth speed and disconnect the upper from the lower cable. Measure the distance lower selector cable projects from cover; and record the measurement for use when reassembling. Disconnect the selector lamp wire and lift off the selector assembly.

To disconnect the pto control cable on tractors so equipped, pull the control handle all the way out, loosen the conduit nut on the transmission cover and the conduit lock nut on the tractor hood (see Fig. FO176). Rotate the control handle clockwise to thread the flexible cable out of the pto connector in the transmission. Remove the lower selector cable from the wheel on the control valve by turning it clockwise to thread it out of the top cover. Remove the ten cap screws securing the top cover to the transmission housing and lift the top cover sufficiently to disconnect the wire

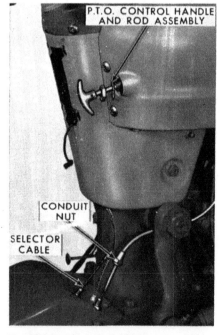

Fig. FO176—Removal of PTO control lever prior to removing transmission top cover.

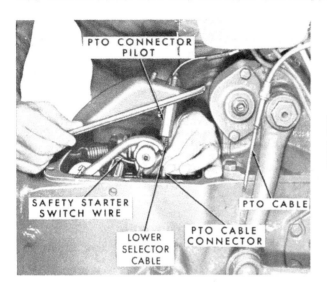

Fig. FO177 — Top cover installation using lower selector cable as a guide for proper positioning of PTO cable connector in PTO connector pilot.

and check the operation of the selector assembly, as outlined in paragraph 228.

When the engine can be started, check adjustment of the PTO control cable as follows: Pull control knob out to mid (feathering) position; the transmission lubrication indicator light should flash on. (If not equipped with indicator light, refer to "NOTE" in paragraph 225, and install pressure gage at rear support.) When the control knob is at the fully in (disengaged) position or at the fully out (engaged) position, the indicator light should go out. Then, slowly move the control knob toward the mid position from both the fully in and fully out positions; travel of control knob before indicator light flashes on should be approximately equal. If not, readjust PTO control cable as follows: Loosen the conduit nuts at each end of the control cable housing. Pull the control handle out and remove the lock nut from control handle housing at hood panel. Turn the control knob at least one full turn clockwise if light flashed on at or too close to the fully in position, or at least one full turn counter-clockwise if the light flashed on at or too close to the fully out position. Tighten the conduit nuts and the control housing lock nut; then, recheck adjustment and make further adjustment as necessary.

from the safety switch, then lift off the cover.

To reinstall the transmission top cover first check to make sure that the control cam is in the "Park" position, with the control valves all in the out position and the rocker arms setting even. Thread the lower selector cable into the pto cable connector to act as a guide for the connector while the cover is being lowered into place (Fig. FO177). Place the cover into position above the transmission housing and attach the safety starter wire before lowering the cover completely into place. When the cover is in place, move the cable up and down to be sure the connector does not bind, install the cover bolts but do not tighten. Thread the lower selector cable out of the pto connector and start the pto cable into place by turning the control handle counter-clockwise until the lower end of the conduit contacts the transmission cover fitting. Tighten the upper and lower conduit nuts and check the control handle for ease of operation. Thread the lower selector cable into the control valve wheel until measurement recorded at disassembly is obtained. If the cable binds, shift the transmission top cover until the cable turns freely. NOTE: If cover bolt holes will not align when cables are free, remove cover, then loosen and shift conduits in cover. If bolt holes still do not align, loosen the control valve mounting cap screws (be careful not to loosen the two cap screws holding top valve housing to lower valve housing) and shift the control valve in direction necessary to allow bolt hole alignment of top cover. Tighten the cover bolts and recheck the lower selector cable for freedom of rotation.

If shift selector has been disassembled, turn the cable in or out as required to obtain a length of 2¾ inches from the top of the fitting to the upper side of the notch in the cable connector as shown in Fig. FO178.

Install the selector assembly with the selector lever in tenth speed. Connect the selector lamp wire and connect the two flexible cables together. Move the selector lever to the Park position which should bring the conduit down to contact with the cover fitting. Secure the selector assembly to the hood, tighten the conduit nut

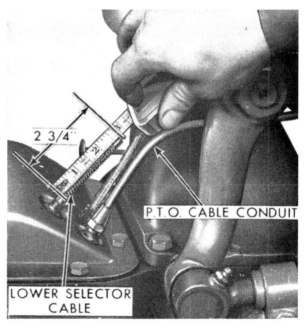

Fig. FO178 — Top cover installation showing measurement for proper positioning of lower selector cable.

235. TRACTOR SPLIT, TRANSMISSION FROM DIFFERENTIAL CENTER HOUSING. To detach (split) transmission from the differential center housing, proceed as follows: Drain the hydraulic reservoir on tractors having a hydraulic system or PTO. Drain transmission as in paragraph 232 if transmission is to be disassembled. Disengage the traction coupling (See paragraph 223) and remove the step plates. Disconnect rear light wire at left side of center housing and pull wire through hole in transmission flange. Disconnect or remove exhaust pipe. On models with transmission lubrication indicator light, remove the inspection (traction coupling shift) plate from left side of center housing and disconnect indicator light wire from sender switch.

On Heavy Duty Industrial tractors, remove the bolts securing side frame members to rear axle housings.

Support both halves of tractor separately and remove bolts and cap screws retaining transmission to differential center housing. Move final drive assembly to the rear away from the transmission housing.

On PTO equipped models, be sure PTO coupling is placed on rear PTO shaft in differential center housing so that it will not be misplaced. Before reattaching transmission to differential center housing, be sure the PTO coupling is on the PTO shaft and the traction coupling sleeve is on the drive shaft with the wide land on the sleeve towards rear.

236. TRACTOR SPLIT, TRANSMISSION FROM ENGINE. Proceed as outlined in paragraph 161, 162 or 163 to attach or detach. Drain the transmission fluid if transmission front cover or inspection plates on either side of housing are to be removed.

237. TRANSMISSION R&R. To remove the transmission, first detach (split) transmission from engine as outlined in paragraphs 161, 162 or 163. Remove the step plates, if not previously removed. On tractors equipped with transmission lubrication indicator light, remove the inspection (traction coupling shift) plate on left side of center housing and disconnect indicator light wire from sender switch. Support transmission separately from final drive assembly and unbolt and remove transmission from the differential center housing. Reverse removal procedure to reinstall transmission.

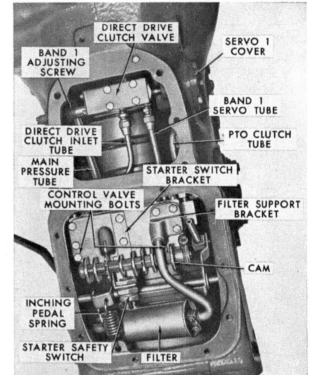

Fig. FO179 — View of late production (1963 & up) transmission with steering gear housing and top cover removed.

238. R&R TRANSMISSION CONTROL VALVE. First, remove transmission top cover as outlined in paragraph 234; then, proceed as follows:

Remove the three cap screws retaining the transmission oil filter tube manifold to control valve and swivel tube back out of way. Remove the two cap screws retaining starter grounding switch bracket to control valve and remove the switch. Remove the two cap screws at left side of control valve. CAUTION: Do not loosen or remove the two remaining cap screws (Control Valve Mounting Bolts—Fig. FO179) in top of control valve body; they retain the valve upper body to valve lower body. Depress the inching pedal (or, preferably, remove the inching pedal return spring) and lift the control valve from top of oil distributor. It may be necessary to pry the valve loose from the gasket seal.

To reinstall control valve, proceed as follows: Place a new gasket on the oil distributor. (Coating the gasket with transmission lubricating oil will aid in holding the gasket in place.) Depress the inching pedal (if pedal return spring has not been removed), place valve in position on gasket and loosely install the two cap screws in left side of control valve. Place new oil filter tube gasket on control valve, swivel oil tube forward and loosely

install the two long cap screws in right side of control valve and the short cap screw in left side of oil tube manifold. Turn the control valve cam to Park position (lobe on starter switch cam pointing to rear), place starter switch and bracket assembly in position and loosely install the two bracket retaining cap screws. Push the switch forward until lobe on cam fully depresses the starter switch plunger, center the switch plunger on the cam lobe and tighten the two bracket retaining cap screws to 5-8 Ft.-Lbs. torque. Then, tighten the remaining cap screws to a torque of 5-8 Ft.-Lbs. Note: Use of proper torque values in tightening valve retaining cap screws is very important. Overtightening will cause binding of control valve spools and improper valve operation. Undertightening will allow pressure leakage.

NOTE: If regulating and pressure relief valves are to be adjusted, it is not necessary to install the starter switch and bracket. Install the two cap screws, however, and tighten them to proper torque value.

239. OVERHAUL TRANSMISSION CONTROL VALVE. Cover shop bench with clean paper to prevent picking up dirt particles when servicing the transmission control valve. Remove the two cap screws retaining upper valve body to lower valve body and separate the two halves. Remove the

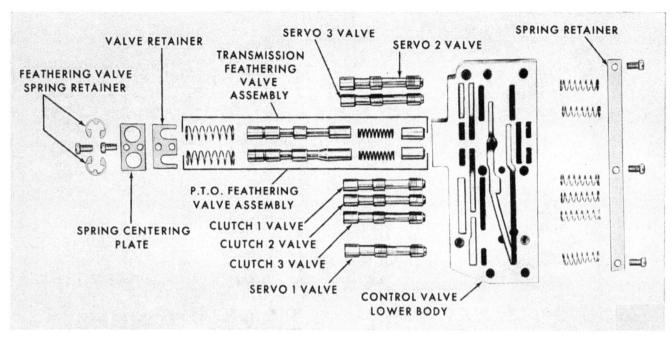

Fig. FO180 — Exploded view of the lower control valve body showing the control and feathering valves, springs and retainers. Note the tapered shoulder on land of PTO feathering valve.

Fig. FO181 — Using the inching pedal to assist in control valve installation if inching pedal return spring has not been removed.

Fig. FO182 — Pulling the cam shaft trunnions for cam shaft removal. Socket shown is not required if 2-inch bolt is used as puller at that end also.

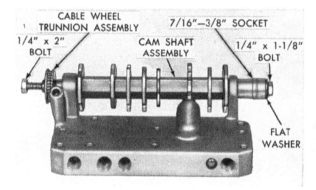

Fig. FO180 for disassembled view of lower body and to Fig. FO183 for disassembled view of upper body.

Check the valve camshaft, detent plunger, cam followers and cam follower shaft for excessive wear or damage. If necessary to renew the camshaft, cable wheel, detent plunger and spring or the upper valve body, proceed as follows: Use a 2-inch ¼ x 20 hardened steel bolt with at least 1¾-inches threaded as a puller. Holding the hex camshaft with a wrench, turn the bolt into the cable wheel to remove wheel and trunnion from end of camshaft. Use the same procedure to remove trunnion from opposite end of camshaft. CAUTION: Do not use a soft or re-threaded bolt for a puller; this type of bolt will twist off in the cable wheel or trunnion. The camshaft, detent plunger and spring can now be removed from the upper valve body. To reassemble, place spring and

detent plunger in bore, position camshaft so that cam with fourteen notches is against detent plunger and insert a ⅜-inch bolt through trunnion bore in body and camshaft. (The bolt must bottom in the camshaft.) Press cable wheel and trunnion assembly into opposite end of camshaft as shown in Fig. FO184 until there is 0.015 clearance between cable wheel and valve body. Thread the puller bolt into cable wheel until it bottoms, then press trunnion into opposite end of camshaft. If necessary to renew any cam followers, cam follower shaft or upper body, remove the snap rings from cam follower shaft and remove the followers, spacers and shaft from body. The cam followers are all identical; however, the spacers must be reinstalled as shown in Fig. FO183 to align the cam followers with the six cams.

The spool valves, feathering valves

six control valve spools and spool return springs from lower body. Remove the feathering valve retaining plate, springs, feathering valves, springs and plungers. The control valve spring retainer can be removed to provide better access for cleaning control spool bores. Remove the adjusting screw retainers, adjusting screws, springs and valves from the rear face of the upper body. Refer to

93

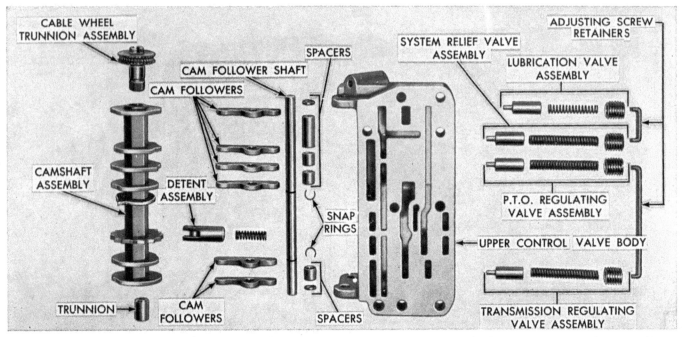

Fig. FO183 — Exploded view of the upper control valve body with the four regulating valves, springs and retainers and the cam shaft and followers.

and regulating valves must all work freely in their bores in the valve body. If the valves or bores are not deeply scored, small burrs can be removed with crocus cloth. Be careful not to round off any sharp edges on the valve spools and thoroughly clean the parts before reassembly. Inspect the upper valve body to be sure all the steel balls used to seal pressure passages are in place. Check the valves, valve bores and springs against the following values:

Valve bore diameter....0.3751-0.3758

Control valve diameter..0.3738-0.3742

Feathering valve
diameter 0.3743-0.3747

Spring Free Length:

Feathering plunger 1.22

Feathering return 1.110-1.130

Valve spool 1.24

Detent 0.70

Lubrication valve 1.47

Relief & regulating valve......2.06

Drop the four regulating and pressure relief valves, hollow end out, in their bores in the upper valve body. Drop the short (1.47) spring in the bore next to cable wheel, and the other (2.06 long) springs in the other three bores. Screw the retainers in flush with valve body. Note: On transmissions without PTO, a pin is used instead of the spring in the PTO regulating valve bore (third bore away from cable wheel end). Tighten the retaining screw against the pin. Do

not install the adjusting screw retainers at this time.

Install the spring retainer plate on back side of lower valve body. Drop the two feathering valve plungers, open end out, in their bores and insert the feathering plunger springs in the plungers. Insert feathering valve with two identifying notches in outer end and tapered shoulder on inner land in the bore towards the side of valve body having four control valve bores. Insert the remaining feathering valve in remaining bore, then install the return springs, snap rings, spring centering plate and feathering valve retainer plate. Drop the six control spool return springs and control spools in their bores. Note: Lubricate the valves in transmission lubricating oil and insert them with a twisting motion to avoid shaving bores with sharp edges of the valves.

NOTE: PTO feathering valve, plunger and springs are not used on transmissions without power take off. Leave this valve bore open when reassembling control valve for these transmissions.

Position the two valve bodies together and loosely install the two retaining cap screws. Holding the two bodies with ends flush and the alignment marks on each end in exact alignment, tighten the two cap screws to a torque of 5-8 Ft.-Lbs. Note: If a new valve body has been installed, turn the camshaft to Park position (all cam followers aligned) and locate

lower valve body to upper valve body so that there is 0.047-0.067 clearance between the side of the discharge ports and the shoulders on the outer (Band 1 and Band 2) valve spools as shown in Fig. FO185. The clearance at each valve spool should be equal. Tighten the two retaining cap screws to a torque of 5-8 Ft.-Lbs. while holding valve bodies in this position. Scribe alignment marks on each end of the two valve bodies.

240. TORQUE LIMITING CLUTCH. To remove the torque limiting clutch, first split the tractor between engine and transmission as outlined in paragraph 165, 166 or 167. Then, unbolt and remove the clutch assembly from the flywheel.

Renew the clutch disc facings (3— Fig. FO186) or clutch disc assembly (2) if facings are glazed, excessively worn or oil soaked. Thickness of disc with new facings is 0.333-0.347.

Renew the Belleville (spring) washer (6) if discolored, cracked or dish measures less than 0.205. Dish in new washer (free height) should measure 0.205-0.215.

Check the splines in the clutch disc hub and on the transmission input shaft for rust or excessive wear. Renew the clutch disc assembly if backlash of disc on input shaft, measured at outer diameter of disc, exceeds ½-inch. Renew the input shaft if backlash of new disc on input shaft

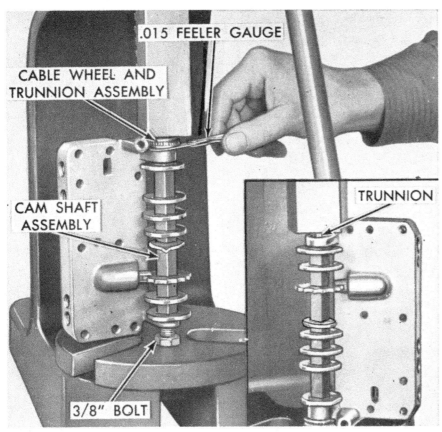

Fig. FO184 — Camshaft installation using a press to install trunnions. Suggested clearance of 0.015 should be maintained between cable wheel and valve body.

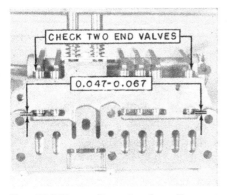

Fig. FO185 — Control valve alignment using feeler gages suggested for proper valve operation.

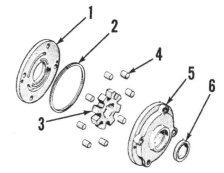

Fig. FO187 — Exploded view of transmission pump assembly. "O" ring (2) and oil seal (6) are only component parts serviced separately from complete pump assembly.

1. Cover	4. Rollers
2. "O" ring	5. Pump body
3. Rotor	6. Oil seal

split the tractor between the transmission and engine assembly as outlined in paragraphs 161, 162 or 163 and unbolt and remove pump. Refer to Fig. FO187 for exploded view of pump.

The roller vane type hydraulic pump can only be renewed as an assembly, however the shaft seal (6) and pump cover "O" ring (2) are serviced separately. The seal can be renewed from the outside without disassembling pump. Use OTC step plate 630-5 or its equivalent to drive seal into pump housing. To renew the "O" ring, remove the two screws retaining the back cover to the pump body, remove the cover, and lift out the "O" ring with a pointed tool. Renew the pump if any of the parts are damaged, scored or excessively worn. Pay particular attention to cam surface in pump body (5). Always renew the mounting gasket when reinstalling the pump. Tighten pump mounting bolts evenly to a torque of 15-18 Ft.-Lbs.

tween this surface and the friction surface at 2.026-2.032.

When reassembling, be sure that concave (cup) side of Belleville washer is towards pressure plate. The hub of the clutch disc is piloted in bore of flywheel; thus, no aligning pilot tool is required. Tighten the clutch housing retaining cap screws to a torque of 25-30 Ft.-Lbs.

241. **TRANSMISSION PUMP.** To remove the transmission pump first

exceeds ¼-inch. Before reassembly, be sure splines in disc hub and on input shaft are clean and free of rust; then, apply a thin film of light silicone grease (Ford part No. M1C-43) to splines of both parts.

Check the friction surfaces of the flywheel and clutch pressure plate for scoring, excessive wear or warping. If flywheel is reconditioned, remove material from clutch housing mounting surface to maintain distance be-

Fig. FO186 — Exploded view of the torque limiting clutch assembly.

1. Flywheel
2. Clutch disc
3. Facings
4. Rivets
5. Pressure plate
6. Belleville washer
7. Clutch cover

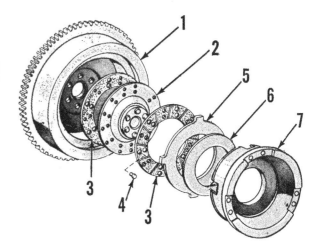
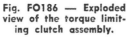

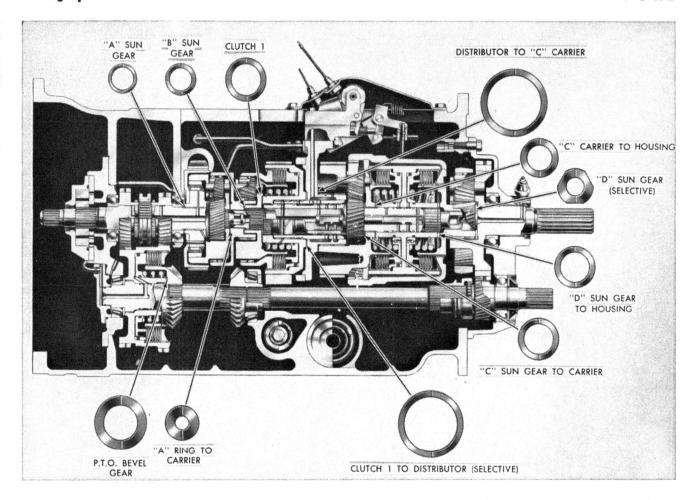

Fig. FO188 — View showing location of thrust washers on 1959-62 production transmissions. Clutch 1 to distributor washer and D sun gear washer are serviced in selective thickness for transmission end play adjustment.

242. TRANSMISSION SHAFT END PLAY. Transmission shaft end play is controlled by bronze thrust washers placed between each of the rotating members. The position and relative size of the thrust washers are shown in Fig. FO188 for 1959-62 production transmissions and in Fig. FO189 for 1963 and later production transmissions. Cumulative wear of all of the thrust washers and wear on thrust surfaces of all transmission components will show up as end play in the transmission. Variation in total length of the components is compensated for during assembly by providing two selective fit washers (Clutch 1 to distributor, and "D" sun gear, Fig. FO188 or Fig. FO189), one at the rear of each of the transmission halves, to hold the end play within specified limits. Cumulative wear on all of the washers will require that this end play be checked and corrected at each overhaul, and renewal of the selective fit washers with one of a greater thickness will provide a correction for wear. Front (Clutch 1 to distributor) washer is available in thicknesses of 0.062 to 0.122 in steps of 0.010. Rear ("D" sun gear to "D" carrier) washer is available in thicknesses of 0.092 to 0.152 in steps of 0.010. End play should be checked as outlined in paragraphs 243 and 244 before disassembly of the transmission components. The overhaul procedures outlined in this manual are based on the supposition that the two selective fit washers are the only ones which will need to be renewed, however, if any of the other thrust washers are other parts in the transmission must be renewed, the additional thickness must be taken into account when determining the thickness of the selective fit washers to be installed.

NOTE: End play of PTO shaft (bearing preload) is important when servicing transmission. On Single-Speed PTO transmissions, excessive PTO shaft bearing preload can cause difficulty in obtaining proper sealing of gasket between transmission case and rear support. Refer to paragraph 261 for PTO shaft bearing preload adjustment procedure.

243. TRANSMISSION FRONT END PLAY. To check transmission front end play, first loosen lock nut and back off Band 2 adjusting screw to release the Band. Then, remove right hand inspection cover as shown in Fig. FO191 and proceed as follows: Pry "A" ring gear rearward with screwdriver and check clearance between "A" ring gear and "B" carrier with feeler gage. Then, pry "A" ring gear forward with screwdriver inserted between "A" ring gear and "B" carrier and again check clearance between these two units with feeler gage. Difference between the two clearance measurements is transmission front end play which should be 0.005-0.015.

NOTE: If transmission is removed from tractor or steering gear is removed from

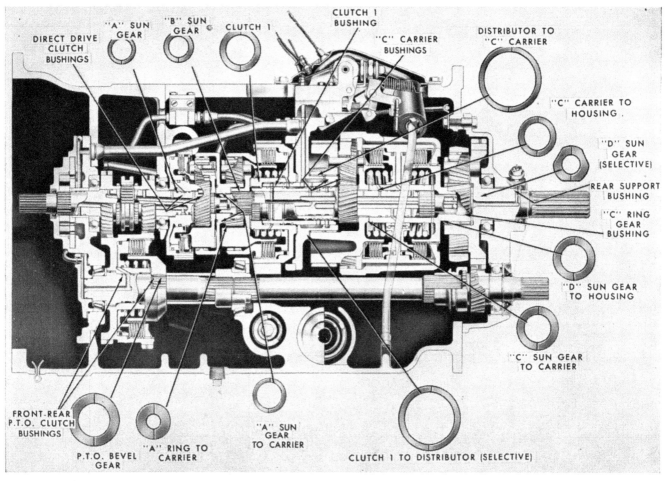

Fig. FO189 — Cross-sectional view showing thrust washer location in late (1963 & up) production transmissions.

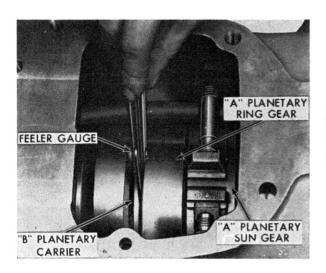

Fig. FO190 — Checking transmission front end play with a feeler gage. End play should be checked prior to disassembly for proper renewal of thrust washers.

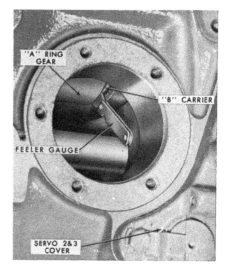

Fig. FO191 — View showing alternate method (See Fig. FO190) of checking transmission front end play if steering gear housing is not removed from transmission.

transmission, front end play can be checked through top opening as shown in Fig. FO190.

244. **TRANSMISSION REAR END PLAY.** After splitting the tractor as outlined in paragraph 235 and before removing any of the transmission components, back off the Band 3 adjusting screw to completely release the band. Mount a dial indicator on the rear of the transmission so that the contact button rests on the output shaft as shown in Fig. FO192. Push the output shaft forward and zero the indicator; then, pull the shaft rearward and measure the end play. Rear transmission end play should be 0.005 to 0.015.

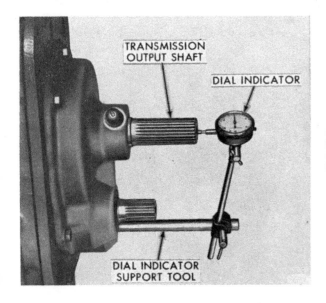

Fig. FO192 — Checking transmission rear end play with a dial indicator prior to transmission disassembly.

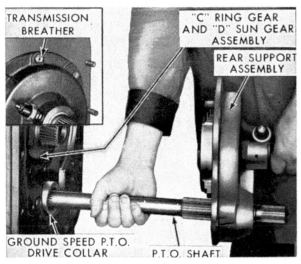

Fig. FO193 — Rear support removal on 2-Speed PTO transmission. PTO shaft will remain in transmission on Single-Speed PTO models.

the vent baffle from the nut and close the vent hole with a 19/32-inch expansion plug. Reinstall the nut on rear support retaining bolt. Install a vent (Ford part No. LA-33592-A or 21T-4022) in top cover near the selector cable housing adapters and reinstall top cover.

The integral transmission output shaft and "D" carrier assembly and the "D" ring gear can be withdrawn from the front side of the rear support after support is off. For removal of the 2-Speed PTO rear shaft and ground speed driven gear, refer to paragraph 260. On 2-Speed PTO transmissions, remove the shifter coupling from the ground speed shifter fork and pivot fork out of way. The "D" sun gear and "C" ring gear assembly can now be removed from transmission. Be careful not to lose the thrust washer located between "D" sun gear and the Clutch 2 and 3 housing.

Except for the ball bearing (10—Fig. FO194), the "D" carrier and transmission output shaft is serviced only as a complete assembly. Inspect the bushing journal on output shaft, the "D" carrier pinions, the thrust washer (13) contact surface and, on 2-Speed PTO transmissions, the ground speed PTO drive gear; renew the "D" carrier assembly if excessive wear or other defects are noted. Inspect the output shaft bushing and the oil seal in the rear support. Renew bushing if worn or scored. Inside diameter of new bushing is 1.749-1.750. Bushing is presized and should not require reaming if carefully installed. Install new seal with lip towards bushing. The "D" sun gear can be separated from the "C" ring gear by removing snap ring (14). Bushing (15) in "D" sun gear is renewable. Inside diameter of new bushing is 0.999-1.000. Bushing is pre-sized and should not require reaming if carefully installed.

NOTE: On 1963 and later production transmissions, the transmission front and rear covers (pump adapter plate and rear support) are line bored after assembly to transmission case; therefore, the rear support is not available for service separately from the transmission case. On 1959-62 production transmissions, the rear support is available separately.

If transmission end play was excessive when measured before disassembly, and no new parts which could change end play are being installed, measure thickness of thrust washer (13) and renew with one of

245. R&R REAR SUPPORT AND "D" PLANETARY SYSTEM. To remove the rear support and "D" planetary system (includes transmission output shaft), first split the tractor between transmission and differential center housing as outlined in paragraph 235 and remove the top cover as outlined in paragraph 234. Check the transmission rear end play as outlined in paragraph 244 and proceed as follows:

First, remove the nut and lockwasher from the top center rear support retaining bolt; then, remove the bolt and the four cap screws from rear support. On 1963 and later production transmissions, screw two ⅜-inch, 24NF thread jackscrews in the tapped holes in rear support and tighten the screws evenly to break the gasket seal. On 1959-62 production transmissions, pry the rear support loose from gasket seal. Then, by pulling on out-

put shaft, withdraw the rear support, "D" carrier and "D" ring gear as a unit as shown in Fig. FO193. On 2-Speed PTO transmissions, the rear PTO shaft and PTO ground drive gears will be removed with the rear support. On Single-Speed PTO transmissions, the PTO shaft will remain in place and must be removed from front of transmission.

NOTE: In early production units, transmission was vented through hollow rear support top center retaining bolt (Transmission Breather—Fig. FO193). Late production transmissions are vented through special vent in top cover to prevent transfer of transmission lubricant to hydraulic sump during "up-hill" operation. If oil transfer through hollow vent bolt is suspected on early production transmissions, the following procedure is recommended: If not already removed, remove top cover as outlined in paragraph 234. Unscrew the nut from rear support top center retaining bolt, remove

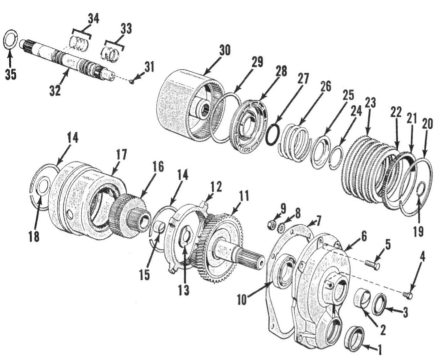

Fig. FO194 — Exploded view of rear support, output shaft, "D" planetary, Clutches 2 and 3 and the transmission main shaft. Plug (1) is used on models without PTO.

1. Plug	9. Lock nut (breather)	17. "C" ring gear	26. Clutch spring
2. Bushing	10. Ball bearing	18. Thrust washer	27. "O" ring
3. Oil seal	11. "D" carrier & trans.	19. Snap ring	28. Clutch piston
4. Plug (lube sender	output shaft	20. Snap ring	29. Piston seal ring
switch)	12. "D" ring gear	21. Pressure plate	30. Clutch 2 & 3 hub
5. Dowel bolt	13. Thrust washer	22. Bronze discs	31. Plug
6. Rear support	14. Snap rings	23. Steel discs	32. Trans. mainshaft
7. Gasket	15. Bushing	24. Snap ring	33. Seal rings
8. Lock washer	16. "D" sun gear	25. Spring retainer	34. Seal rings
			35. Thrust washer

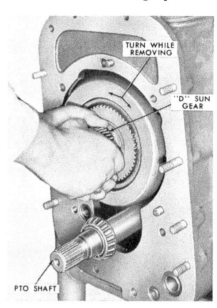

Fig. FO195 — Removing "D" sun gear — "C" ring gear assembly from Single Speed PTO transmission.

Fig. FO196 — Removal of main shaft, "C" carrier and Clutches 2 and 3.

proper thickness to reduce end play to 0.005-0.015.

To reassemble, proceed as follows: Using Lubriplate, stick thrust washer (18) to Clutch 2 and 3 thrust surface, being sure recess in the thrust washer is over snap ring (19). Install "C" ring gear and "D" sun gear as an assembly. Rotate the assembly back and forth while applying light forward pressure, to align splines with Clutch 3 discs and "C" ring gear with "C" carrier pinions.

Lubricate the bushing and oil seal in rear support and insert the output shaft and "D" carrier assembly. Fit the "D" ring gear over the "D" carrier pinions and into the retaining notches in rear support; then, using a new gasket, install the assembly. On 2-Speed PTO transmissions, the ground speed shift collar must be in place in the shifter fork and the rear PTO shaft inserted through the collar. On Single-Speed PTO transmissions, carefully guide the PTO shaft through the oil seal in rear support, install retaining cap screws and top center

retaining bolt loosely; then, tighten the lower cap screws at each side of the PTO shaft first. On all models, tighten the cap screws and top center bolt to a torque of 35-40 Ft.-Lbs. Note: When inserting top center retaining bolt on 2-Speed PTO transmissions, be sure the bolt enters the PTO interlock cable housing support bracket. After tightening the retaining cap screws and bolt, install the lock washer and nut on top center bolt.

246. MAINSHAFT, CLUTCH 2 AND 3 ASSEMBLY, "C" CARRIER, "C" SUN GEAR AND BAND 3. After removal of the rear support, "D" planetary and "C" ring gear as outlined in paragraph 245, the mainshaft, "C" carrier and the Clutch 2 and 3 assembly can be removed as a unit. Back off the Band 3 adjusting screw until Band 3 is loose, then withdraw the mainshaft, "C" carrier and Clutch 2 and 3 assembly from rear of transmission as shown in Fig. FO196. Compress Band 3 and remove band and struts from transmission (Fig. FO197). Carefully withdraw "C" sun gear

from distributor (Fig. FO198). Remove thrust washer from rear face of distributor.

Remove the four cast iron sealing rings (34—Fig. FO194) from front end of mainshaft and slide the "C" carrier and two thrust washers off front end of shaft. Remove snap ring at rear end of shaft, withdraw shaft from Clutch 2 and 3 assembly and remove the three cast iron sealing rings from rear end of shaft.

To remove Clutch 3 components from housing, proceed as follows: Remove snap ring (20), pressure plate (21) and lift the five bronze plates (22) and five steel plates (23) from housing. Using Nuday tool No. N-775, place clutch housing in press as shown in Fig. FO199 and depress spring seat (25—Fig. FO194) far enough to remove snap ring (24). Be sure that

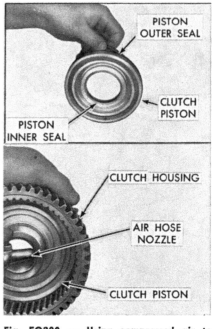

Fig. FO197 — Band 3 with adjusting and actuating mechanism. Adjusting and actuating struts are secured only by band, actuating lever and adjusting screw.

Fig. FO198 — Removing "C" sun gear by withdrawing it from distributor. Seals are cast iron sealing rings.

Fig. FO199 — To remove clutch piston, compress spring with a suitable straddle mounted tool and remove snap ring.

Fig. FO200 — Using compressed air to force clutch piston from housing. Piston seals are "O"-ring type.

Fig. FO201 — Checking the coning of the steel clutch plates with a feeler gage. Plates should be renewed if coned less than 0.015.

spring seat does not enter snap ring groove while releasing press. Remove tool, spring seat and spring. Use air pressure (Fig. FO200) in Clutch 3 port inside clutch housing hub to force piston out of clutch housing. Remove seal ring (29—Fig. FO194) from outer diameter of piston and remove "O" ring (27) from inner diameter of piston. Clutch 2 components can be removed from front side of clutch housing in similar manner. Clutch 2 and Clutch 3 components are identical.

Renew the die-cast piston if warped, scored or otherwise damaged. Install new seal ring on outer diameter and new "O" ring in bore of piston. Be sure the clutch housing and passageways are clean, lubricate piston sealing rings with Lubriplate and press piston, flat side in, into clutch housing. Place spring and spring seat over piston, depress spring seat in press using Nuday tool No. N-775 and install retaining snap ring. Release spring pressure against seat and snap ring.

The steel clutch discs are coned (cupped) 0.015-0.020. Check steel discs against flat surface as shown in Fig. FO201 and renew any with less than 0.015 clearance at inner diameter. Also renew any steel disc that is warped or scored.

The bronze clutch discs are flat. Renew any bronze disc that is not flat or is burned, scored or excessively worn.

When reinstalling clutch discs, a steel disc must be installed next to the piston and the bronze and steel discs alternately placed. Also, the concave (cupped) side of the steel discs must be towards the pressure plate. It is not necessary to align the wide notches in outer edge of the steel plates. Check the pressure plate for scoring, warpage, excessive wear or cracks. Install the pressure plate and retaining snap ring.

Except for the front two bushings, the "C" carrier is serviced as a complete assembly only. If renewal of bushings is indicated, remove old bushings with cape chisel as shown in Fig. FO202. Install new bushings with Bushing Adapter, Nuday tool No. 777. Bushings are pre-sized and should not require reaming if carefully installed. Inside diameter of new bushings is 3.0615-3.0625. Check to see that "C" carrier pinions are in good condition and that pinion shafts are tight in the carrier. Also check end thrust surfaces on the carrier; renew carrier as an assembly if defect other than worn or scored bushings is noted.

NOTE: Early production "C" carriers also contained a rear bushing which was lubricated from a drilled hole in the mainshaft. Late "C" carriers do not have this bushing, and carrier to mainshaft clearance is approximately 0.015. If installing a new type "C" carrier without bushing, either renew the mainshaft also or restrict the lubrication hole by pressing a roll pin, Ford part No. 305019-S, into the mainshaft drilled hole so that end of pin is flush with shaft.

Renew Band 3 if friction material inside band is worn or band has been overheated. On 1959-62 production transmissions, Band 3 is identical with Band 2, the bands having a non-metallic friction lining. On 1963 and later production transmissions, Band 2 has a metallic friction lining and Band 3 has a semi-metallic friction lining. The bands are dimensionally identical; however, the band with a metallic lining should not be used in the Band 3 position. The late type bands can be installed in early pro-

duction transmission. If both bands are to be removed from transmission, tag Band 3 so that it may be reinstalled in the correct location. When reinstalling Band 3, be sure that end of flat strut having notch is towards the band and that the identifying notch in the band is towards the band adjusting screw. Install thrust washer (21—Fig. FO211) on distributor.

NOTE: Following paragraph 247 outlines procedure for reinstalling the "C" sun gear, mainshaft, "C" carrier and the Clutch 2 and 3 assembly as a unit. For alternate procedure of installing the "C" sun gear, the mainshaft and "C" carrier and the Clutch 2 and 3 assembly as separate units, as preferred by some mechanics, refer to paragraph 248.

247. Install three new seal rings on rear end of mainshaft, lubricate rings with Lubriplate, center the ring end gaps on top of shaft and insert mainshaft through the Clutch 2 and 3 assembly as shown in Fig. FO203. Insert snap ring in groove at rear end of mainshaft. Stand the assembly on rear end of shaft, place thrust washer (35 —Fig. FO194) over front end of shaft against clutch hub and install "C" carrier on shaft as shown in Fig. FO-204. Rotate the carrier back and forth to align splines with Clutch 2 discs. When carrier contacts thrust washer, install front thrust washer (20—Fig. FO211) on shaft against carrier and install four new sealing rings on front end of shaft. Lubricate the seal rings with Lubriplate and align ring end gaps at top side of shaft. Install six new sealing rings (23) on

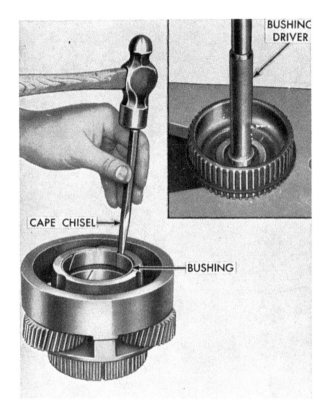

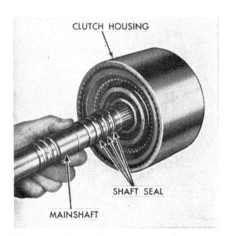

Fig. FO203 — Installing mainshaft into clutch housing. Mainshaft serves as fluid distributor line for the three clutches.

Fig. FO202 — Bushing removal. Design of "C" carrier prevents removal of front bushings with a press. Remove with a cape chisel as shown and install with press and a suitable adapter.

Fig. FO204 — Installing "C" carrier and "C" sun gear on main shaft and clutch housing. Fluid for the three clutches must pass through the "C" sun gear shaft.

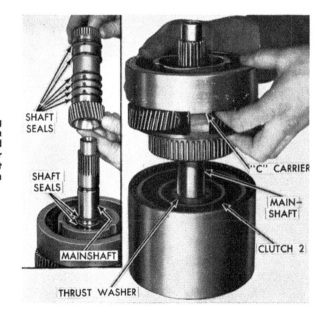

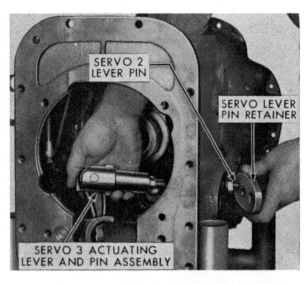

Fig. FO205 — Removing Servo 3 actuating pin and lever assembly. Pin retainer prevents pins from rotating in transmission housing.

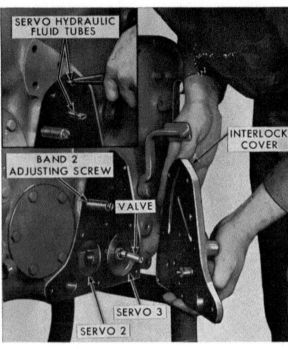

Fig. FO206 — Removing the interlock cover to pull the servo fluid tubes.

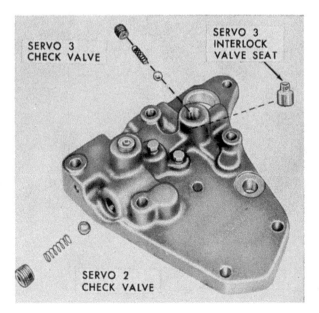

Fig. FO207 — Exploded view of transmission interlock cover. Refer to Fig. FO208 for correct installation of Servo 3 interlock valve seat.

"C" sun gear and carefully slide sun gear over the front end of mainshaft. Lubricate the sealing rings on sun gear with Lubriplate, align the ring end gaps at top side of shaft and, holding the "C" carrier against Clutch 2 and 3 assembly, carefully install the "C" sun gear through the distributor. Rotate mainshaft back and forth slightly to align splines with "B" carrier splines and, holding pinion gear stationary, rotate "C" carrier back and forth slightly to align "C" sun gear splines with splines in Clutch 1 housing.

248. As an alternate method from that outlined in paragraph 247, some mechanics prefer to install the "C" sun gear, mainshaft, "C" carrier and the Clutch 2 and 3 assembly as follows:

Install six new sealing rings (23—Fig. FO211) on "C" sun gear, lubricate rings with Lubriplate, align ring end gaps at top of shaft and carefully insert sun gear in distributor. Rotate sun gear slightly to align with Clutch 1 housing splines.

Place thrust washer (35—Fig. FO-194) over front end of mainshaft. Insert the mainshaft through "C" carrier, place thrust washer (20—Fig. FO211) over front end of shaft and slide it against carrier. Install four new sealing rings on front end of mainshaft, lubricate the sealing rings with Lubriplate and align the ring end gaps. Holding shaft with ring end gaps up, insert shaft through "C" sun gear, carefully working the shaft forward to avoid breaking rings. Rotate mainshaft slightly to align splines with "B" carrier. Tighten Band 3 adjusting screw when "C" carrier is in place.

Install three new sealing rings on rear end of mainshaft, lubricate the rings with Lubriplate, align ring end gaps at top of shaft and carefully install the Clutch 2 and 3 assembly on shaft to avoid breaking sealing rings. Partly support the clutch housing and rotate housing slightly from side to side while applying light forward pressure to align Clutch 2 discs with splines on "C" carrier. When housing is in position against the thrust washer and "C" carrier, pull the mainshaft rearward while supporting clutch housing, so that snap ring (19) can be installed in notch at rear end of mainshaft.

249. **INTERLOCK COVER.** The interlock cover assembly can be removed from the left side of transmis-

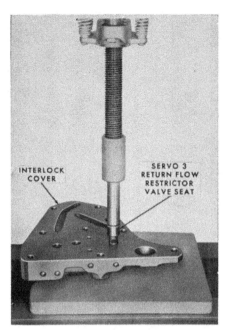

Fig. FO208 — Servo 3 interlock valve seat must be reinstalled with cross-hole in seat at right angle to bottom (straight) side of the interlock cover as shown.

interlock cover. Note: The new Servo 3 interlock valve seat must be reinstalled with the 5/16-inch hole in a vertical position (at right angle to the straight side of interlock cover) as shown in Fig. FO208.

Before reinstalling the interlock cover, be sure that gasket surfaces of cover and transmission case are clean, and remove any burrs that may have been caused when prying cover loose. Check to see that the interlock valve plunger on Servo 3 shaft is securely attached and the straight pin is in place in the Servo 3 interlock valve seat. Then, using a new gasket, carefully install cover over the valve plunger on Servo 3. Tighten the cover retaining cap screws to a torque of 35-40 Ft.-Lbs.

250. **DISTRIBUTOR, CLUTCH 1 AND "B" CARRIER.** The oil distributor, Clutch 1 assembly and the "B"

carrier can be removed after removing mainshaft and "C" sun gear as outlined in paragraph 246 and the interlock cover as outlined in paragraph 249. Check transmission front end play as outlined in paragraph 243, remove control valve as outlined in paragraph 238 (if not already removed); then, proceed as follows:

Remove Servo 2 and 3 actuating pin retaining nuts, washers, retaining plate and "O" rings from right side of transmission case. Back the Band 3 adjusting screw out as far as possible. On Single-Speed PTO transmissions, push the Servo 3 actuating pin and lever against side of transmission. On transmissions without power take-off or with 2-Speed PTO, remove Servo 3 actuating pin and lever assembly.

On PTO equipped transmissions, disconnect fitting on PTO clutch pressure line at front right corner of dis-

sion case without removing other transmission components; however, it must be removed to allow removal of the oil distributor plate as outlined in paragraph 250. To remove the interlock cover, proceed as follows:

If removing the interlock cover without removing other transmission components, drive pin out of inching pedal and pivot the pedal to rear out of way.

If top cover has been removed, remove the inching pedal return spring and disconnect the inching pedal control rod at feathering valve bellcrank.

Remove the lock nut and sealing washer from Band 2 adjusting screw; then, back screw out until it is free. Remove the eight interlock cover retaining screws and pry cover loose from gasket seal. Refer to Fig. FO206.

CAUTION: The interlock cover must be removed evenly to avoid breaking the interlock valve that is attached to the No. 3 Servo shaft. Also, be careful not to lose the 3/16-inch, 1.38 long straight pin from the Servo 3 interlock valve seat in interlock cover.

The Servo 2 and Servo 3 check valves (0.469 diameter steel balls) and springs can be removed from interlock cover after removing the socket head retaining plugs. Refer to exploded view of interlock cover in Fig. FO207. After removing the Servo 3 check valve, the check valve seat can be pressed from inner side of

Fig. FO209 — Removing the distributor, Clutch 1 and "B" planetary. Note position of ground speed PTO shifter fork.

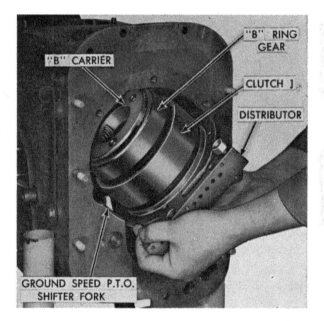

Fig. FO210 — Removing distributor from Single-Speed PTO transmission with PTO shaft in place.

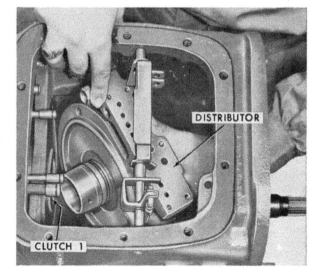

tributor. Loosen the four distributor retaining cap screws and, using a pair of pliers, pull Servo 2 and 3 pressure tubes from left side of transmission case as shown in Fig. FO206. Remove the distributor retaining cap screws. On Single-Speed PTO transmissions, insert a long screwdriver through center of distributor against splines in Clutch 1 housing and hold the clutch in place while prying oil distributor rearward. Tip top left corner of distributor out to clear Servo 3 actuating pin and lever, then remove distributor from transmission as shown in Fig. FO210. Then, pull Clutch 1 assembly and the "B" carrier out to rear of transmission.

NOTE: On transmissions without power take-off or on 2-Speed PTO transmissions, the distributor, Clutch 1 assembly and "B" carrier can be removed as a unit as shown in Fig. FO209.

Fig. FO211 shows an exploded view of the distributor, "C" carrier, Band 3 and related parts. The "B" carrier, Clutch 1 assembly, Band 2 and related parts are shown in the exploded view in Fig. FO212.

The "B" carrier is serviced as a complete assembly only. Planetary "B" ring gear also serves as the Clutch 1 pressure plate. After removing the "B" ring gear, service procedures for Clutch 1 are the same as outlined for Clutch 3 in paragraph 246. Renew the bushing (15—Fig. FO212) if scored or excessively worn. Inside diameter of new bushing is 2.2178-2.2188. Bushing is pre-sized and should not require reaming if carefully installed.

Inspect the upper face of distributor for burrs, scratches and flatness and check the oil passages for obstructions. Carefully examine both the inner and outer wearing surfaces of distributor center tube and the end thrust surfaces at front and rear sides of distributor plate for burrs, scoring or excessive wear. Renew the distributor if any of these defects are noted. Install new "O" rings on rear end of main supply tube, Servo 1 pressure tube and, on 1963 and later production, on rear end of Direct Drive Clutch pressure tube.

To reassemble, place "B" carrier on work bench, pinion side down, and stick thrust washer (3—Fig. FO212) in counterbore of carrier with Lubriplate. Place the assembled Clutch 1 and "B" ring gear unit, ring gear side down, over the "B" carrier. While partially supporting the clutch, slight-

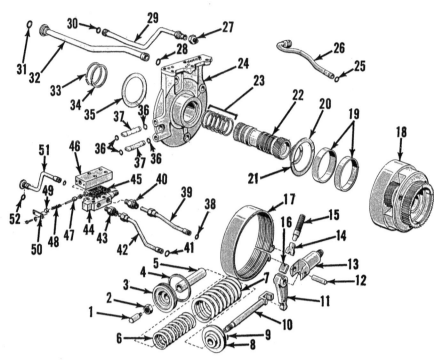

Fig. FO211 — Exploded view of distributor, "C" carrier, "C" sun gear, No. 3 band and related parts. Servo 1 oil tube (26) is used on 1959-62 production transmissions; on 1963 & later transmissions, Direct Drive Clutch valve, oil pressure tube and related parts (items 38 through 52) are used.

1. Interlock valve	15. Adjusting screw	29. PTO clutch tube	42. Direct drive clutch
2. Nut	16. Strut	30. "O" ring	tube
3. Servo 3 piston	17. Band 3	31. "O" ring	43. Fitting
4. "O" ring	18. "C" carrier	32. Main supply tube	44. Valve support
5. Sleeve	19. Bushings	33. Seal ring	45. Gasket
6. Servo spring	20. Thrust washer	34. Seal ring	46. Valve body
7. Servo spring	21. Thrust washer	(stepped ends)	47. Direct drive clutch
8. Spring retainer	22. "C" sun gear	35. Thrust washer	valve
9. Roll pin	23. Seal rings	36. "O" rings	48. Spring
10. Servo piston rod	24. Distributor	37. Servo 2 & 3 tubes	49. Plug
11. Actuating lever	25. "O" ring	38. "O" rings	50. Retainer plate
12. Groove pin	26. Servo 1 tube	39. Servo 1 tube	51. Direct drive clutch
13. Actuating pin	27. Fitting	40. Fitting	tube
14. Strut	28. "O" ring	41. "O" ring	52. "O" ring

ly rotate the assembly back and forth to align the clutch discs with the splines on "B" carrier and the "B" ring gear with the carrier pinions. Be sure that thrust washer (1) is in place in hub of "B" sun gear; then, holding Clutch 1 and "B" carrier together, install them on the "B" sun gear.

If transmission front end play was incorrect when measured before disassembly as outlined in paragraph 242, and no new parts were installed which could change end play, measure old thrust washer (35—Fig. FO-211) and renew with one of proper thickness to provide 0.005-0.015 end play. Place thrust washer on front side of distributor and install new seal rings on distributor with the ring having the locking ends in the rear groove. Lubricate the sealing rings on distributor and the "O" rings on rear end of main supply and pressure tubes with Lubriplate and align ring end gaps at top of distributor. Insert

the front end of the distributor through the Clutch 1 hub, carefully compress front seal ring with screwdrivers and push distributor into place. It will be necessary to align the main supply tube, Servo 1 pressure tube, Direct Drive Clutch pressure tube (1963 and later production) and the PTO clutch pressure tube (PTO equipped models) with the distributor as it is being pushed into place. It may be necessary to lift distributor up slightly to engage pilot diameter of distributor in bore of transmission case.

Loosely install the four distributor retaining cap screws. Insert the Servo 2 and 3 pressure tubes through left side of transmission case; then, install new "O" rings on the tubes. Lubricate the "O" rings and push the tubes into bores in distributor. Be sure that the beveled openings in tubes are aligned with oil passage slots in interlock cover. Note: Gasket marks can usually be seen on case, or hold gasket up to transmission as a guide.

Tighten the distributor retaining cap screws to a torque of 20-25 Ft.-Lbs. and recheck transmission front end play as outlined in paragraph 242. If end play is not within 0.005-0.015, remove the distributor and renew thrust washer (35) with one of proper thickness to provide correct transmission front end play. Connect and tighten the PTO clutch pressure line fitting.

Reinstall the Band 3 actuating lever and pin assembly, if removed. Install new "O" rings on the Band 2 and 3 actuating lever brackets, install retainer plate and loosely install the retainer plate cap screw. Install the flat washers and nuts on the lever brackets and tighten the nuts to a torque of 100-120 Ft.-Lbs. Tighten the retainer plate cap screw to a torque of 90-110 Ft.-Lbs.

251. "A" RING GEAR, "A" CARRIER AND OVERRUNNING CLUTCH. (1959-62 production transmissions only; refer to paragraph 252 for 1963 and later transmissions.) After removing the distributor, Clutch 1 and "B" carrier as outlined in paragraph 250, the "A" ring gear ("B" sun gear), "A" carrier and overrunning clutch can be withdrawn from the transmission case as shown in Fig. FO213. An exploded view of these parts is shown in Fig. FO214.

The "A" carrier is serviced as a complete assembly only. Carefully inspect the pinions and the thrust surfaces and renew "A" carrier if defects are noted.

Bushings in the "A" sun gear and drum assembly are renewable; however, other components are serviced as a complete assembly only. Inside diameter of new bushings is 1.249-1.250. Bushings are pre-sized and should not require reaming if carefully installed.

Place the "A" sun gear and drum front end down on a bench, with "A" carrier still in place. The "A" carrier assembly should turn freely in a counter-clockwise direction, but should lock tightly to "A" sun gear when turned clockwise. To disassemble for inspection or renewal of the one-way roller clutch or other components, turn the "A" carrier counterclockwise while withdrawing from sun gear assembly.

Remove the snap ring retaining the one-way clutch to "A" sun gear; then lift out the clutch assembly as a unit

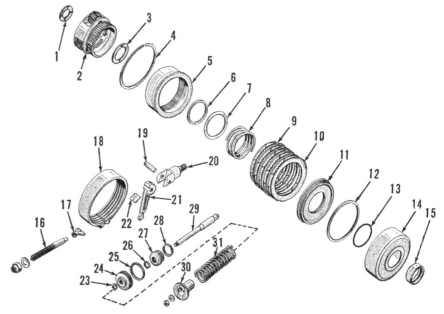

Fig. FO212 — Exploded view of "B" planetary, Clutch 1, Band 2 and related parts.

1. Thrust washer	9. Bronze discs	17. Strut	25. "O" ring
2. "B" carrier	10. Steel discs	18. Band 2	26. "O" ring
3. Thrust washer	11. Clutch piston	19. Groove pin	27. Guide
4. Snap ring	12. Piston seal ring	20. Actuating pin	28. "O" ring
5. "B" sun gear	13. "O" ring	21. Actuating lever	29. Servo piston rod
6. Snap ring	14. Clutch 1 housing	22. Strut	30. Spring retainer
7. Spring retainer	15. Bushing	23. Snap ring	31. Servo spring
8. Clutch spring	16. Adjusting screw	24. Servo 2 piston	

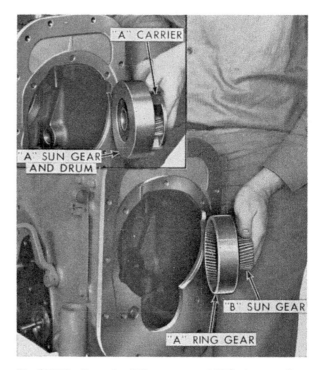

Fig. FO213—Removing "B" sun gear and "A" planetary from transmission case on 1959-62 production transmission.

while applying slight counter-clockwise pressure. When assembling the one-way clutch, refer to Fig. FO215 and Fig. FO216. The upper end bearing and spring are crimped to the roller cage and should not be removed. A steel spacer (9—Fig. FO-214) is located behind the roller clutch assembly.

To reassemble the lower end bearing and spring, hook offset end of spring (2—Fig. FO215) in bearing cage as shown; and the other end in lug (4) of end bearing (3). Rotate end bearing (3) counter-clockwise until bearing is properly positioned over roller cage and spring (2) is fully wound inside the end bearing.

Be sure spacer (9—Fig. FO214) is in place on "A" sun gear hub; then, invert the assembled roller unit as shown in Fig. FO216, and install over the inner race; holding the lower end bearing in position while installation is being made. The arrow or crimped end bearing should face upward as shown. The "crimped boss" lugs on ID of upper and lower end bearings should fit in adjacent valleys of inner race. Do not attempt to align the lugs. After installation, the roller cage should have approximately ¼-inch free play on the sun gear. Install the snap ring as shown in Fig. FO217. Thread one end of the ring in the groove, holding the offset and opposite end of the ring out of the groove with 0.015 feeler gage (A). Lock the upper free end of the ring to the center portion by lining up and inserting the stamped projection into the elongated slot (B). With the drum firmly held, remove the feeler gage and snap the rest of the ring into the groove using a screwdriver (C). Recheck the roller assembly for free movement; then turn the roller cage counter-clockwise to take up all slack. Install

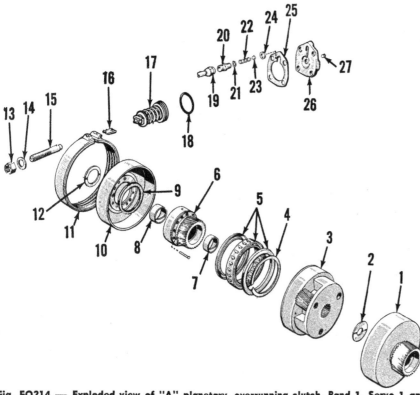

Fig. FO214 — Exploded view of "A" planetary, overrunning clutch, Band 1, Servo 1 and related parts used on 1959-62 production transmissions. Remove and discard items 21 through 24 if found. Refer to Fig. FO219 for 1963 & later production transmissions.

1. "A" ring gear — "B" sun gear	6. "A" sun gear	13. Lock nut
2. Thrust washer	7. Bushing	14. Sealing washer
3. "A" carrier	8. Bushing	15. Adjusting screw
4. Snap ring	9. Spacer	16. Strut
5. Overrunning clutch assy.	10. Brake drum	17. Servo 1 assy.
	11. Band 1	18. "O" ring
	12. Thrust washer	19. Servo 1 tube
		20. Fitting
21. Flat washer		
22. Spring		
23. Check valve		
24. Valve seat		
25. Gasket		
26. Servo 1 cover		
27. Plug		

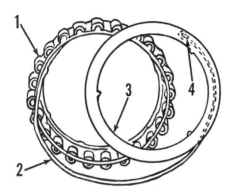

Fig. FO215 — Assembling the one-way clutch. Refer to text.

1. Roller assembly
2. Spring
3. End bearing
4. Lug

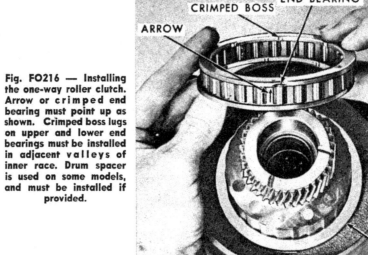

Fig. FO216 — Installing the one-way roller clutch. Arrow or crimped end bearing must point up as shown. Crimped boss lugs on upper and lower end bearings must be installed in adjacent valleys of inner race. Drum spacer is used on some models, and must be installed if provided.

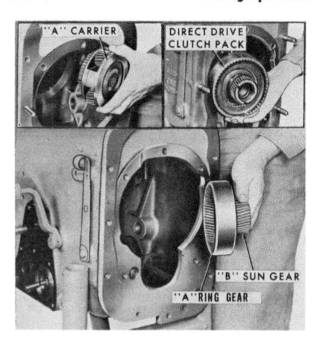

Fig. FO218 — Removing "A" planetary unit and the Direct Drive Clutch assembly from transmission.

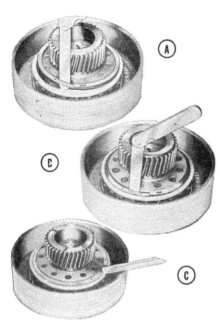

Fig. FO217 — Installing the interlocking type snap ring on the one-way clutch. See text.

"A" carrier by rotating counter-clockwise while lowering in position; then check to see that carrier rotates freely when turned counter-clockwise, but immediately locks up when turned clockwise.

After being sure thrust washer (12 —Fig. FO214) is in place on input shaft, install "A" sun gear, overrunning clutch and "A" carrier as a unit. Install thrust washer (2) on input shaft, sticking the washer in place with Lubriplate, and then install "A" ring gear ("B" sun gear) (1).

252. **"A" RING GEAR, "A" CARRIER AND DIRECT DRIVE CLUTCH.** (1963 and later production only; refer to paragraph 251 for 1959-62 production transmissions.) After the distributor, Clutch 1 assembly and "B" carrier have been removed as outlined in paragraph 250, the "A" ring gear ("B" sun gear), the "A" carrier and the Direct Drive Clutch assembly can be withdrawn from the transmission as shown in Fig. FO218.

To disassemble and overhaul the Direct Drive Clutch assembly, proceed as follows: Remove the large snap ring from inside diameter of clutch housing and lift out the pressure plate, two bronze and two steel clutch discs. Place the spring com-

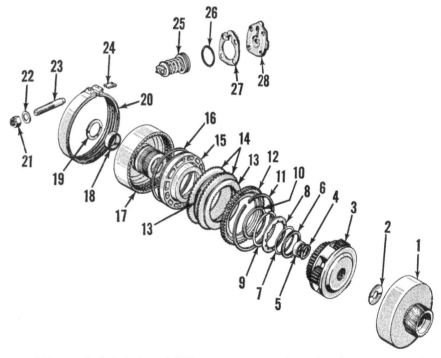

Fig. FO219 — Exploded view of "A" planetary, Direct Drive Clutch assembly, Band 1 and Servo 1 as used in 1963 and later production transmissions.

1. "A" ring gear — "B" sun gear	8. Spring washer	15. Clutch piston	22. Sealing washer
2. Thrust washer	9. Pivot ring	16. Piston seal ring	23. Adjusting screw
3. "A" carrier	10. "O" ring	17. Clutch housing	24. Strut
4. Thrust washer	11. Snap ring	18. Bushing	25. Servo 1 assy.
5. Bushing	12. Pressure plate	19. Thrust washer	26. "O" ring
6. Snap ring	13. Bronze discs	20. Band 1	27. Gasket
7. Pivot ring	14. Steel discs	21. Lock nut	28. Servo cover

pressor (Nuday tool No. N-488) shown in Fig. FO220 inside the clutch housing and reinstall the large snap ring, making sure the ends of snap ring are not positioned over slot of compressor tool.

NOTE: Compressor tool has three stepped, slotted studs spaced equidistantly around its outer edge. The steps are cut in increments of 0.006 and are used in compressing the Belleville (spring) washer (8—Fig. FO219) to allow removal of the snap ring (6).

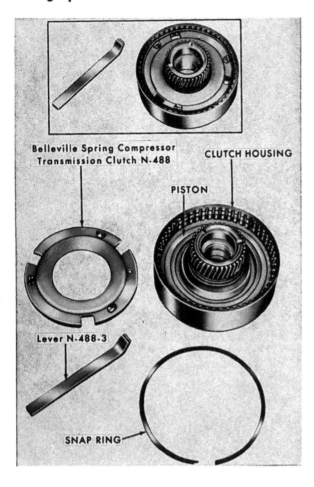

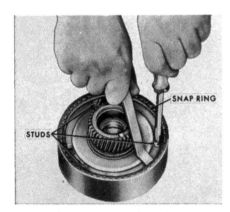

Fig. FO220 — Special tools used to disassemble the Direct Drive Clutch assembly.

Fig. FO221—Compress Direct Drive Clutch spring washer as shown using special tools shown in Fig. FO220. Refer to text.

Use lever (N-488-3) and screwdriver as shown in Fig. FO221 and depress compressor tool so that first step of each stud can be engaged under the snap ring. Repeat this operation for the remaining two steps of the three studs. When the third step of each stud is under the snap ring, remove snap ring (6—Fig. FO219) from hub of clutch housing. Remove the compressor tool and lift the small pivot ring (7), Belleville (spring) washer (8) and large pivot ring (9) from clutch hub. Remove the clutch piston (15) and renew the "O" ring (10) and piston outer seal (16). Coat the "O" ring and seal with Lubriplate and carefully reinstall piston in housing after making sure that housing is clean.

Bushings (5 and 18) in the Direct Drive Clutch housing are renewable. Inside diameter of new front bushing is 1.374-1.375; inside diameter of new rear bushing is 1.249-1.250. Bushings are pre-sized and should not require reaming if carefully installed.

The two steel plates, the two bronze plates and the friction surface of the Direct Drive Clutch are flat. Renew the plates if they are warped, scored or excessively worn. Renew the 0.036 thick Belleville washer if cracked or if free height measures less than 0.115.

Reassemble the Direct Drive Clutch by reversing disassembly procedure. Place a steel disc with either side towards piston, install a bronze disc, the second steel disc and then install the second bronze disc. Then, reinstall pressure plate and retaining snap ring.

The "A" carrier is renewable as a complete assembly only. Check pinions and thrust surfaces and renew the carrier if defect is noted.

Check the two sealing rings (17—Fig. FO227) and the thrust washer (19—Fig. FO219) on rear end of transmission input shaft and renew if worn, scored or broken. Place new thrust washer against the input shaft rear bearing, then install the lock-end sealing rings in the two grooves in input shaft. Lubricate the sealing rings with Lubriplate and align ring end gaps at top of shaft.

Place the "A" carrier, with splines up, on work bench. Stick the thrust washer (4) in recess in carrier hub. Set the assembled Direct Drive Clutch on the carrier and rotate the clutch housing back and forth to align clutch discs with splines on carrier. Holding the carrier and clutch together, install them as a unit taking care to work the clutch housing over the sealing rings on transmission input shaft.

253. **DIRECT DRIVE CLUTCH VALVE. (1963 and later production transmissions only.)** The Direct Drive Clutch control valve can be unbolted and removed from its support after the steering gear housing is removed from transmission, when the tractor is split between the engine and transmission as outlined in paragraph 236, or when the transmission is removed. The valve support can be removed after removing the front cover (pump adapter plate) as outlined in paragraph 257. Remove the Direct Drive Clutch front pressure tube from front of transmission case, disconnect the rear clutch pressure tube and Servo 1 pressure tube from rear side of valve support, remove the fittings and unscrew support from transmission case.

Disassemble Direct Drive Clutch control valve by removing the two screws, retainer plate, stop plate, spring and valve from valve body. Clean all parts and inspect valve and valve bore for scoring, nicks, burrs or other damage. Valve should slide freely in its bore. Inspect spring for distortion or possible fractures. Check to see that all oil passages are clear. Refer to Fig. FO211 for exploded view of valve assembly.

Renew "O" rings on clutch pressure tubes and Servo 1 pressure tube when reassembling.

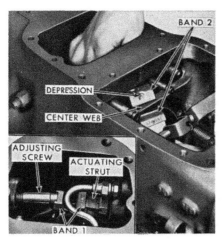

Fig. FO222 — Removing Band 2 by rotating band around transmission center web. Inset shows Band 1 in position.

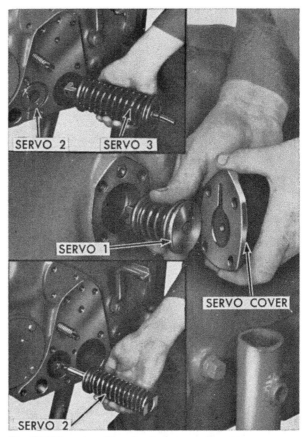

Fig. FO223 — Installing the three servo assemblies. Note the position of the actuating notches on Servos 2 and 3.

254. BANDS 1 AND 2, STRUTS AND ACTUATING LINKAGE. After removing the "A" carrier and over-running clutch as outlined in paragraph 251 or the "A" carrier and Direct Drive Clutch as outlined in paragraph 252, Bands 1 and 2 can be removed as follows: Loosen Band 1 adjusting screw, compress Band 1 and remove the band and actuating strut out rear of transmission case. Compress Band 2 and remove the two actuating struts. To remove Band 2, rotate it around center web of transmission case as shown in Fig. FO222.

On transmissions with Single-Speed PTO, Band 2 actuating lever and pin assembly cannot be removed until the PTO shaft is removed from the transmission as outlined in paragraph 259.

NOTE: On 1959-62 production transmissions, Bands 2 and 3 had non-metallic friction lining and were interchangeable. On 1963 and later production transmissions, Band 2 has metallic friction lining and Band 3 has a semi-metallic friction lining. Although the new type bands have the same dimensions, only the band with a semi-metallic lining should be used in the Band 3 location. Therefore, tag Band 3 before removing Band 2. The new type bands can be installed in early transmissions.

255. RENEW SERVO PISTON SEALS. Servo seals can be renewed without disassembling the transmission if care is used not to dislodge the loose band struts. To renew the servo seals, proceed as follows:

SERVO 1: Refer to Fig. FO223 and remove the four cover attaching bolts and Servo 1 cover. Back off the lock nut on Band 1 adjusting screw (Fig. FO169) while holding the adjusting screw stationary; then, carefully tighten the adjusting screw until the servo piston is forced from its bore just enough to expose the "O" ring seal. Apply slight pressure on the servo piston while renewing the seal to avoid dislodging the strut. Back off the adjusting screw while applying pressure on the servo piston, reinstall the cover and tighten the retaining cap screws to 20-25 Ft.-Lbs. Readjust the band as outlined in paragraph 227.

NOTE: If necessary to remove Servo 1, or Band 1 strut is dropped into transmission while renewing Servo 1 seal ring, drain the transmission and proceed as follows: Remove the inspection cover from right side of transmission case, locate and remove the Band 1 strut. Remove Servo 1 from transmission, use a small amount of heavy grease to stick strut to servo shaft and carefully reinstall the servo in transmission.

SERVO 2: Drain the transmission as outlined in paragraph 232, completely loosen Band 2 and Band 3 adjusting screws and remove left step plate, exhaust pipe and the interlock cover. Remove the right step plate, Servo 2 and 3 actuating pin nuts and flat washers, and retainer bolt and retainer. Unbolt and remove the inspection cover from right side of transmission case for access during reassembly. Tighten Band 2 adjusting screw until servo spring is forced from transmission housing enough that it can be grasped firmly by one hand, rotate spring either way 180 degrees to disengage notch in servo piston rod from actuating lever and carefully withdraw servo assembly from housing. Servo guide (27—Fig. FO212) will usually be removed with servo assembly; if not, extract guide with a suitable hooked tool and renew "O" ring seals on guide and servo piston. Reinstall servo guide over piston rod and insert the assembly in the transmission housing with the notch forward. Feeling through the removed inspection cover, be sure the notched end of the servo piston is in line with the lower end of actuating lever (21—Fig. FO212); then, rotate outer end of actuating pin with a wrench until the milled flat is vertical and the actuating pin retainer can be reinstalled. Tighten actuating pin retaining nuts to a torque of 100-120 Ft.-Lbs. Care must be taken not to move actuating lever inward to dislodge the actuating strut while removing and installing the servo. Reassemble the remainder of the trans-

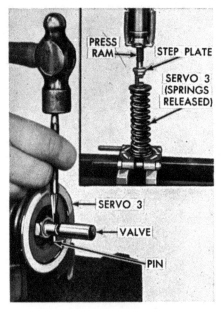

Fig. FO224 — Disassembling Servo 3. Always remove the interlock valve piston before disassembly of the servo.

Fig. FO225 — View showing method of removing transmission front cover (pump adapter plate) on early transmissions without tapped jack screw holes.

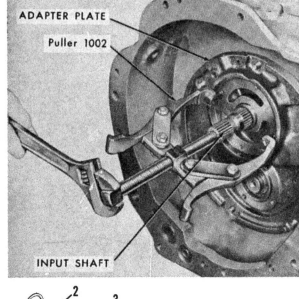

Fig. FO226 — Exploded view of transmission front cover and alternate input shaft assemblies used on 1959-62 production transmissions. Shaft (3) is used on Single Speed PTO transmissions; shaft (13) is used on 2-Speed PTO transmissions. Plugs (6 & 10) are used on transmissions not equipped with power take-off.

1. Snap ring	5. Gasket	9. Pump assembly	13. Input shaft
2. Ball bearing	6. Plug	10. Plug	14. Shift collar
3. Input shaft	7. Pump adapter plate	11. "O" ring	15. 1000 RPM drive
4. Ball bearing	8. Gasket	12. 540 RPM drive gear	gear

mission parts, refill the transmission and adjust the bands as outlined in paragraph 227.

SERVO 3: Drain the transmission as outlined in paragraph 232, completely loosen Band 2 and 3 adjusting screws and remove the left step plate, exhaust pipe and interlock cover. Grasp the interlock valve attached to the end of Servo 3 and withdraw the servo to expose the sealing "O" ring. If sufficient slack is not available to expose the seal, loosen the adjusting screw while withdrawing the servo. Renew the "O" ring while being careful not to force the piston back in the housing bore. Tighten the adjusting screw to draw the servo piston into its bore, then reinstall the interlock plate and the remainder of the parts. Fill the transmission and adjust the bands as outlined in paragraph 227.

256. R&R AND OVERHAUL SERVO ASSEMBLIES. Servos 1 and 2 can be removed as outlined in paragraph 255. To remove Servo 3, first remove the mainshaft, "C" carrier and Clutch 2 and 3 assembly as outlined in paragraph 246 and remove the interlock cover as outlined in paragraph 249. Then remove Servo 3 from transmission.

Servo 1 is serviced as a complete assembly only except for the "O" ring on the piston. Note: On early production transmissions, a ball type restrictor valve (items 21, 22, 23 and 24—Fig. FO214) was located in a bore in transmission case just above the

Servo 1 piston bore. This restrictor valve should be removed and discarded, where present, as follows: Insert a punch which is a tight fit in the valve seat (24) and wiggle seat out of case. Remove the steel ball (23), spring (22) and flat washer (21) and discard all parts (21 through 24).

If renewal of Servo 2 or 3 piston rod, piston, springs or retainers is necessary, disassemble servo as follows: On Servo 3, drive the pin out of the servo rod as shown in Fig. FO224 and remove the interlock valve piston. On either servo, compress the spring in a suitable press as shown in Fig. FO224 and remove the retaining nut. Release spring tension and disassemble the servo. Reassemble after renewing damaged or worn parts by reversing the disassembly procedure. Note: Be sure that the Servo 2 piston retaining snap ring (23—Fig. FO212) is installed with the sharp edge **away** from the piston.

257. TRANSMISSION FRONT COVER. To remove the transmission front cover (pump adapter plate), the transmission must be detached from engine as outlined in paragraph 161, 162 or 163. Then, proceed as follows:

Remove the PTO pressure tube from front cover and PTO front bearing retainer. Then, unbolt and remove PTO front bearing retainer, taking care not to lose or damage any of the shims between retainer and front cover. Unbolt and remove the transmission pump. It may be necessary to rap the pump body a glancing blow with a soft faced hammer to break the pump gasket seal.

On 1959-62 production transmissions, screw two cap screws (**do not** use pump retaining cap screws) into front cover and pull the cover loose with pullers against input shaft as shown in Fig. FO225. Usually, considerable force is required to break the gasket seal.

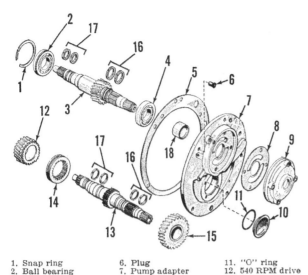

Fig. FO227 — Exploded view of transmission front cover (pump adapter plate) and alternate input shaft assemblies used on 1963 and later production transmissions. Refer to Fig. FO226 for components parts of earlier production transmissions.

1. Snap ring
2. Ball bearing
3. Input shaft
4. Ball bearing
5. Gasket
6. Plug
7. Pump adapter plate
9. Pump assembly
10. Plug
11. "O" ring
12. 540 RPM drive gear
13. Input shaft
14. Shift collar
15. 1000 RPM drive gear
16. Sealing rings
17. Sealing rings
18. Sleeve

Fig. FO228 — Removing the adapter plate and input shaft.

On 1963 and later production transmissions, screw two ⅜-inch, 24NF thread, jack screws into the tapped holes at each side of front cover. Turn the jack screws in equally to pull the front cover loose from sealing gasket.

CAUTION: Do not allow transmission input shaft to move forward while removing front cover, or the thrust washer (2—Fig. FO214 or Fig. FO219) between the "A" carrier and "A" ring gear may drop out of place. If this occurs, it will be necessary to remove all of the transmission components from rear of transmission so that the thrust washer can be reinstalled.

At this time, the main supply tube, PTO pressure tube, Servo 1 tube and, on 1963 and later production transmissions, the Direct Drive Clutch control valve, tubes and valve support can be removed.

The 2-Speed PTO front shaft can now be removed and reinstalled as follows: Pry between washer and PTO clutch housing to pull shaft and front bearing forward. Removing inspection plate from side of transmission case will aid in shaft removal. Install the two cast iron sealing rings on shaft, lubricate the seals with Lubriplate, align seal ring end gaps on top of shaft and insert shaft through the PTO clutch housing while holding splined hub in clutch. Working through the inspection cover opening, install spacer (30—Fig. FO231), shims (29) if present, and rear bearing adapter (31) on shaft as it is inserted through clutch housing.

On 1963 and later production transmissions, check the sealing ring seating surface in sleeve (18—Fig. FO227) located in front cover. If worn or scored, renew by pressing damaged sleeve out. Press new sleeve in so that end of sleeve having chamfered inside diameter is flush with recessed machined surface at rear side of front cover. Install new sealing rings on front end of input shaft, lock ring ends together, lubricate rings with Lubriplate and align rings ends on top of shaft. Using new gasket, install cover plate taking care not to break the sealing rings. PTO shaft can be lifted up to enter bore in front cover by inserting pin or screwdriver in front end of shaft.

On 1959-62 production transmissions, install front cover using new gasket. Lift front end of PTO shaft so that it will enter bore in cover plate by installing pin or screwdriver in front end of shaft.

On all models, tighten the front cover retaining nuts to a torque of 25-30 Ft.-Lbs. Install PTO front bearing cup in front cover. Install new sealing ring on front end of PTO shaft, lock ends of ring together and lubricate ring with Lubriplate. Place shims on PTO front bearing retainer and install new "O" ring on retainer. Install retainer on front cover and tighten the retaining cap screws to a torque of 12-15 Ft.-Lbs. Check the PTO shaft bearing preload as outlined in paragraph 261 and add or remove shims as necessary. Using a new gasket, reinstall pump on front cover and tighten retaining cap screws to a torque of 15-18 Ft.-Lbs.

Refer to paragraph 239 concerning

splines on transmission input shaft and in torque limiting clutch disc hub.

NOTE: On 1963 and later production transmissions, the transmission front cover (pump adapter plate) is not available separately from the transmission case and rear support (rear cover). The front and rear covers are line bored and therefore are not interchangeable between different transmission cases. Front cover, rear cover and transmission case are available separately for service on 1959-62 production transmissions.

258. **INPUT SHAFT, BEARINGS AND PTO DRIVE GEARS.** To remove the transmission input shaft, it is necessary to remove the transmission from tractor as outlined in paragraph 237. Then, remove all transmission components from rear of case as outlined in paragraphs 245 through 252 and remove the front cover as outlined in paragraph 257. The input shaft can then be removed out towards front as shown in Fig. FO228.

On 1963 and later production, remove the four cast iron sealing rings (16 and 17—Fig. FO227) from input shaft. On all models, pull the ball bearings from each end of shaft if renewal of shaft, bearings or 2-Speed PTO drive gears and/or sliding coupling is indicated.

On 1959-62 production transmissions, input shaft front bearings with two different outside diameters are used. When obtaining replacement bearing, be sure that outside diameter of new bearing is same as that of removed bearing. If renewing the transmission front cover (pump adapter plate), obtain corresponding input shaft front bearing.

Fig. FO229 — Removing the pto clutch on the 2-speed PTO transmission.

Fig. FO230 — Removing Single Speed PTO transmission shaft. Snap ring (see inset) behind PTO clutch assembly encircles only slightly more than half way around shaft and can be driven off the shaft with punch.

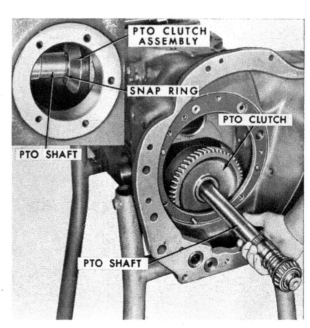

On 2-Speed PTO transmissions, install large PTO drive gear on front end of shaft with clutch teeth towards splined coupling machined on shaft. Then, press front bearing onto shaft with ball filling notch in bearing race towards front end of shaft. Slide the coupling over splined connector and the small PTO drive gear onto rear end of shaft with clutch teeth towards coupling. Then, press rear input shaft bearing on shaft with snap ring groove towards front end of shaft. Be sure both bearings are pressed firmly against shoulders on shaft; press on center race only.

On Single-Speed PTO transmissions, or transmissions without power take-off, install the input shaft bearings as outlined in preceding paragraph.

259. PTO CLUTCH. To remove the PTO clutch, the transmission input shaft must first be removed as outlined in paragraph 258.

On 2-Speed PTO transmissions, the clutch and front PTO shaft can be removed as a unit as shown in Fig. FO229. Disassemble the unit by removing rear bearing adapter and pulling shaft forward out of unit. Be careful not to lose any shim(s) that may be present at either end of the spacer sleeve on shaft. Pull front bearing cone and roller and thrust washer from shaft, and rear bearing cone and roller from rear bearing adapter if renewal is indicated.

To remove the Single-Speed PTO clutch, first remove the bearing cone and roller (29—Fig. FO232), washer (28) and snap ring from rear end of

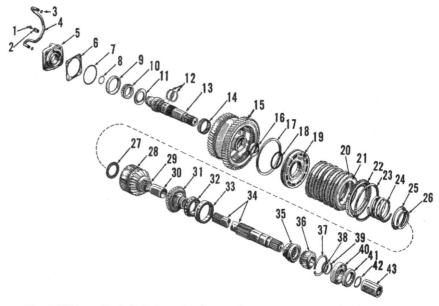

Fig. FO231 — Exploded view of 2-Speed PTO clutch, shaft and related parts.

1. Phillips screw	12. Sealing rings	23. Snap ring	33. Bearing cup
2. Retainer	13. Front PTO shaft	24. Clutch spring	34. Rear PTO shaft
3. "O" rings	14. Bushing	25. Spring retainer	35. Shift collar
4. PTO clutch tube	15. PTO clutch housing	26. Snap ring	36. Ground speed gear
5. Bearing retainer	16. Bushing	27. Thrust washer	37. Snap ring
6. Shims	17. Piston seal	28. Front bevel gear	38. Snap ring
7. "O" ring	18. "O" ring	29. Shims (1959-62 only)	39. Thrust washer
8. Sealing ring	19. Clutch piston	30. Spacer	40. Ball bearing
9. Bearing cup	20. Steel discs	31. Rear bevel gear	41. Seal
10. Cone & roller assy.	21. Bronze discs	(bearing adapter)	42. Snap ring
11. Thrust washer	22. Pressure plate	32. Bearing cone & roller	43. Coupling

shaft. Using a small punch or screwdriver, drive snap ring (25) out of groove on shaft behind the splined clutch hub (24). Then, pull shaft forward out of clutch assembly and lift clutch from transmission.

Bushings (14 and 16—Fig. FO231, or 9 and 12—Fig. FO232) in clutch housing are renewable. Inside diam-

eter of new front bushing is 1.874-1.875; inside diameter of new rear bushing is 1.624-1.625. Bushings are pre-sized and reaming should not be necessary if carefully installed.

Clutch pistons, piston and seals, discs and piston return spring can be serviced as outlined for the Clutch 2 and 3 assembly in paragraph 246. The

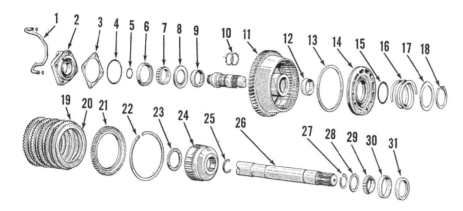

Fig. FO232 — Exploded view of the Single Speed PTO assembly.

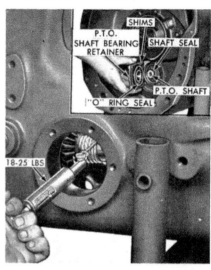

Fig. FO233 — Measuring pto shaft bearing preload. Inset shows location of adjusting shims.

1. Pto front pressure tube	9. Front bushing	16. Clutch spring	24. Clutch hub
2. Bearing retainer	10. Sealing rings	17. Retaining washer	25. Retaining ring
3. Retainer gasket	11. Gear & housing assembly	18. Snap ring	26. Pto front shaft
4. "O" ring	12. Bushing	19. Steel clutch plate	27. Snap ring
5. Oil seal	13. Piston sealing ring	20. Bronze clutch plate	28. Washer
6. Bearing cup	14. Clutch piston	21. Pressure plate	29. Cone & roller assy.
7. Cone & roller assy.	15. Piston sealing ring	22. Snap ring	30. Bearing cup
8. Thrust washer		23. Thrust washer	31. Oil seal

2-Speed PTO clutch has six bronze and six steel discs; the Single-Speed PTO clutch has five bronze and five steel plates.

After clutch is reassembled, stick the thrust washer (27—Fig. FO231 or 23—Fig. FO232) in recess in front side of bevel gear (28—Fig. FO231) of 2-Speed PTO clutch or splined hub (24—Fig. FO232) of Single-Speed PTO clutch, using Lubriplate. Insert the bevel gear or splined hub into clutch discs until thrust washer contacts thrust surface of clutch housing.

On 2-Speed PTO unit, assemble the clutch and shaft on bench as follows: Place thrust washer (11—Fig. FO231) on front end of shaft and press front bearing cone and roller (10) tightly against washer. Install the two sealing rings (12) from rear end of shaft, lubricate the rings with Lubriplate and align ring end gaps. While holding the bevel gear (28) securely in clutch, insert shaft through clutch housing with ring end gaps up. Then, install shim(s) (29) if used, spacer sleeve (30) and rear adapter (31) with rear bearing cone and roller (32) installed. Reinstall unit in transmission.

NOTE: On 1959-62 production 2-Speed PTO transmissions, shim(s) are used between spacer sleeve (30—Fig. FO231) and bevel gear (28) to properly space the two bevel gears (28 and 31). On 1963 and later production transmissions, a heavier spacer (30) is used on which the length is closely controlled; eliminating the need for shims. When servicing a 1959-62 transmission with 2-Speed PTO, take care not to lose any of

the shims present on disassembly and always reinstall the same number of shims. If, for some reason, a late production sleeve (Ford part No. CONN-A786-A) is installed in a 1959-62 production transmission, discard the shims.

On Single-Speed PTO, install thrust washer (8—Fig. FO232) on front end of PTO shaft and press front bearing cone and roller (7) tightly against washer. Install two sealing rings (10) from rear end of shaft, lubricate the seal rings with Lubriplate and align ring end gaps. Holding the splined hub in clutch assembly, place clutch in transmission and insert shaft through clutch with sealing ring end gaps at top of shaft. When thrust washer (8) is against clutch housing, drive the snap ring (25) into groove at rear of splined hub. Install the snap ring (27), washer (28) and bearing cone and roller (30) on rear end of shaft.

260. 2-SPEED PTO GROUND DRIVE SPEED GEARS. All 2-Speed PTO transmissions also have a ground speed PTO ratio, in which the PTO shaft is driven from the transmission output shaft.

The ground drive PTO drive gear is integral with the transmission output shaft and "D" carrier assembly. The driven gear (36—Fig. FO231) is located on the rear transmission PTO shaft.

To remove the ground drive gears, remove the transmission rear support as outlined in paragraph 245. The drive gear is serviced as an assembly

with the "D" carrier only. To remove the driven gear, proceed as follows: Remove the PTO shaft oil seal (41—Fig. FO231) from rear support and then remove snap ring (42) from shaft (34). The PTO shaft can then be bumped forward out of the ball bearing (40). Remove the thrust washer (39) and snap ring (38); then, slide PTO driven gear (36) from rear end of shaft. If necessary to renew bearing (40), first remove snap ring (37) and drive bearing forward out of rear support. Reassemble by reversing disassembly procedure. Be sure seal surface of shaft (34) is not damaged before installing new seal (41).

261. PTO SHAFT END PLAY (BEARING PRELOAD). End play (bearing preload) of the power take-off clutch and shaft unit is controlled by varying the thickness of the shim stack (6—Fig. FO231 or 3—Fig. FO232) between the transmission front cover (pump adapter plate) and the PTO front bearing retainer. Shims are available in thicknesses of 0.002, 0.003, 0.010 and 0.030. Check bearing preload as follows:

On 2-Speed PTO, bearing preload is checked with the rear support and rear PTO shaft removed from the transmission.

On Single-Speed PTO, bearing preload is checked with the PTO shaft oil seal removed from the transmission rear support (cover).

Tighten the PTO front bearing retainer (2) cap screws to a torque of 12-15 Ft.-Lbs. With the inspection

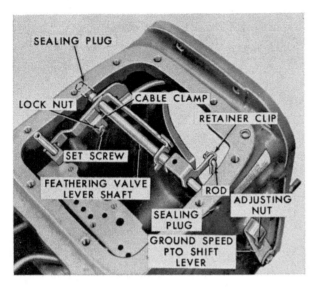

Fig. FO234 — Adjusting PTO interlock cable on 2 - Speed PTO transmissions. Refer to text.

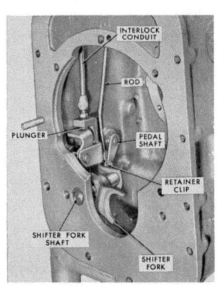

Fig. FO235 — View showing ground speed shift fork and interlock plunger which prevents ground speed PTO from being engaged at same time engine speed PTO is engaged. Refer to text and also to Fig. FO234.

cover on side of transmission removed, wrap a cord around the PTO shaft and attach a pull scale to the cord as shown in Fig. FO233. Bearing preload is correct when a steady pull of 18 to 25 pounds is required to keep the shaft rolling. Add or remove shims as required to obtain proper bearing preload.

262. TRANSMISSION CASE LINKAGE. Refer to Figs. FO234, FO235 and FO236. Usually, the linkage does not need to be removed from the transmission case when overhauling the transmission. However, if it is necessary to renew linkage or sealing "O" rings, refer to the following paragraphs:

INCHING PEDAL AND CONTROL VALVE LINKAGE. To install new inching pedal, drive pin from pedal and remove pedal from shaft. Remove the transmission top cover as outlined in paragraph 234, hold shaft in place with pry bar and install new pedal. If pedal shaft (see Fig. FO235), sealing "O" ring or control rod are to be renewed, "C" carrier, mainshaft and Clutch 2 and 3 assembly must first be removed as outlined in paragraph 246.

NOTE: On 1963 and later production transmissions, a longer inching pedal is used which locates the transmission feathering point closer to the step plate and increases pedal travel through feathering point to improve inching control. The new type pedal can be installed on 1959-62 transmissions. (A different pedal is required on H.D. Industrial tractors than on other models.) After installing the new type pedal, be sure lubrication indicator light goes out when pedal is fully depressed against the step plate.

(On tractors not having an indicator light, install a 0-300 psi gage in Servo 2 as shown in Fig. FO166. Shift transmission to Neutral position and observe gage reading with inching pedal released. Depress pedal to step plate; gage reading should then be as high, or higher, than with pedal released.) If lubrication indicator light does not go out, or low gage reading is observed with pedal in fully depressed position, bend the step plate down so that proper pedal travel can be obtained.

To remove the feathering valve lever shaft (see Fig. FO234) or components located on the shaft, disconnect inching pedal return spring, remove snap rings from shaft and drive against either sealing plug with thin punch. Shaft need not be driven all the way out to renew feathering valve levers or spacers. After reinstalling shaft, insert sealing plugs and check to be sure feathering valve levers are centered against the feathering valves. Relocate shaft by driving against sealing plugs as necessary. Note: On transmissions without power take-off, the PTO feathering valve lever and associated parts are not used.

PTO INTERLOCK LINKAGE. (2-Speed PTO transmissions only.) Interlock linkage is required on 2-Speed PTO transmissions to prevent the ground drive and engine drive PTO systems from both being engaged at the same time and locking up the transmission. Refer to Fig. FO234. The linkage adjustment can be checked as follows: Move the ground speed lever to "OFF" position and pull the PTO control knob on hood panel to fully out (engaged) position. It then should not be possible to move the

ground speed lever to "ON" position. Push the PTO knob on hood panel in and move ground speed lever to "ON" position. It then should not be possible to pull the PTO knob on hood panel to the out (engaged) position.

To adjust the PTO interlock linkage, the transmission top cover must first be removed as outlined in paragraph 234. Note: The interlock linkage can be adjusted or adjustment checked whenever the transmission control valve, ground speed shifting mechanism and interlock linkage are installed. To adjust the linkage, proceed as follows:

Move the ground speed PTO shift lever to the "ON" position. Note: If lever does not engage in detent, loosen adjusting nut on lever and readjust detent position. Pull up on the PTO control connector. There then should be 0.005-0.015 clearance between the end of the PTO feathering valve and the PTO feathering valve lever. If not, loosen the set screw (see Fig. FO234) clamping the interlock cable in the feathering lever trunnion, move the feathering lever until the 0.005-0.015 clearance is obtained and tighten the set screw. Note: Clearance should always be checked when pulling up on connector.

To service the ground speed shift and interlock mechanism, the "C" carrier, mainshaft and Clutch 2 and 3 assembly must be removed as outlined in paragraph 246.

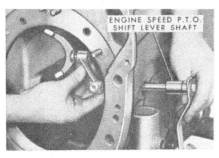

Fig. FO236 — Installing 540-1000 RPM shift lever shaft on 2-Speed PTO transmission.

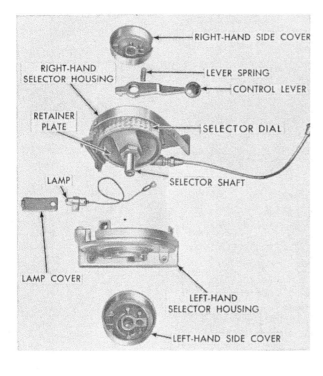

540 AND 1000 RPM SHIFT LINKAGE. The shift linkage can be renewed or a new sealing "O" ring installed after removing the transmission front cover as outlined in paragraph 257. To remove linkage, first remove pin, slide shaft out, and remove arm and fork as shown in Fig. FO236.

262. SELECTOR ASSEMBLY. If service is required on the selector assembly it can be removed for servicing by disconnecting the cable housing and cable at the transmission top cover, disconnecting the lamp wire and unbolting and removing the selector assembly.

To disassemble, first remove the two side covers and the three Phillips head screws securing the left hand selector housing. Remove the hex nut from the shaft and the snap ring and retainer plate to expose the wheel and cable assembly.

NOTE: Shock torque is usually required when removing shaft nut, to prevent twisting shaft.

To assemble, reverse the disassembly procedure. Fig. FO237 and Fig. FO238 show the components of the selector assembly.

Fig. FO237 — Disassembled view of the selector assembly showing component parts.

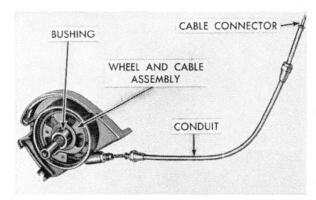

Fig. FO238 — Partially disassembled view of selector wheel and cable assembly.

DIFFERENTIAL, BEVEL GEARS AND AXLE OR FINAL DRIVE

R&R DIFFERENTIAL
Offset Models

265. To remove the differential unit, first drain the rear axle center housing. Before the right rear final drive unit can be removed, it will be necessary to remove the counterweight and hydraulic cover offset housing as follows:

266. R&R COUNTERWEIGHT. Jack up the right rear wheel and remove the wheel and brake drum. Remove the inner bearing retainer (1—Fig. FO240) from the wheel axle shaft. Remove the retainer nut and bearing and drive the wheel axle shaft out of the drop housing. Unbolt and remove the brake backing plate, support the counterweight with a suitable hoist, remove the three bolts (A—Fig. FO-241) securing the counterweight to the axle housing and remove the weight.

After reinstalling the counterweight, adjust the wheel axle shaft bearings as outlined in paragraphs 278 and 279.

267. R&R HYDRAULIC COVER OFFSET HOUSING. Disconnect the hydraulic linkage and unbolt and remove the tractor seat. Disconnect the quadrant linkage and remove the bolts connecting the hydraulic lift cover to the right fender. Unbolt and remove the hydraulic lift manifold, remove the remainder of the bolts securing the lift cover to the offset housing and lift the cover from the housing with a suitable hoist. Unbolt and remove the left fender, unbolt the pto shifter plate and move the pto shift lever outward to clear offset housing. Disconnect the sump return line and the parking brake levers and unbolt and remove the offset housing.

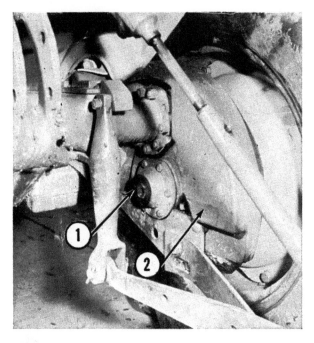

Fig. FO240 — View of right final drive on Offset type tractor, showing (1) axle shaft bearing retainer and (2) 300 lb. counterweight.

Fig. FO242 — Before removing hydraulic cover offset housing, left fender must be unbolted and removed.

Fig. FO241 — To remove counterweight, first remove wheel, wheel axle and brake assembly, then remove three attaching bolts (A).

268. **R&R RIGHT FINAL DRIVE AND DIFFERENTIAL ASSEMBLY.** After removal of the rear axle counterweight as outlined in paragraph 266 and the hydraulic cover offset housing as outlined in paragraph 267, removal of the right final drive housing is conventional. Support the tractor under the rear axle center housing, remove the step plate and unbolt and remove the right final drive housing and long axle shaft. The right rear fender may be removed to prevent damage if preferred. Withdraw the differential unit from the center housing.

When reinstalling the final drive housing, use only one standard gasket between it and the center housing to obtain the correct preload on the differential carrier bearings. Tighten retaining nuts to a torque of 55-65 Ft.-Lbs. prior to serial No. 106387 (1958-62 production), or to 65-70 Ft.-Lbs.

on models with serial No. 106387 and up.

All-Purpose, LCG, Utility Industrial and Grove Types

269. To remove the differential unit, first drain the rear axle center housing. Block up under the center housing to raise the rear wheels, remove the left rear wheel and disconnect the left brake linkage. Unbolt the left axle housing from the center housing and move the housing and axle assembly away from the tractor. Withdraw the differential unit from the center housing.

When reinstalling the axle or final drive unit tractors with $\frac{7}{16}$ inch housing retaining studs use only one standard gasket between it and the center housing to obtain the correct preload on the differential carrier bearings.

Tractors equipped with $\frac{9}{16}$ inch housing retaining studs do not use a gasket, but are equipped with "O" ring housing seals. On models with $\frac{7}{16}$-inch retaining studs, tighten the retaining nuts to 66-65 Ft.-Lbs. prior to serial No. 106387 (1958-62 production), or to 65-70 Ft.-Lbs. on models with serial No. 106387 and up. On all models equipped with $\frac{9}{16}$-inch studs, tighten retaining nuts to 130-150 Ft. Lbs.

Rowcrop Type

270. To remove the differential unit, first drain the rear axle center housing. Block up under the center housing to raise the rear wheels and remove the right rear wheel. Disconnect the right brake linkage and hydraulic link, unbolt the right final

drive housing from the center housing and move the unit away from the tractor. Withdraw the differential unit from the center housing.

When reinstalling the final drive unit, use only one standard gasket between it and the center housing to obtain the correct preload on the differential carrier bearings. Tighten retaining nuts to a torque of 55-65 Ft.-Lbs. prior to serial No. 106387 (1958-62 production), or to 65-70 Ft.-Lbs. on models with serial No. 106387 and up.

Heavy Duty Industrial Type

271. To remove the differential unit, drain the rear axle center housing and block up under the housing to raise the rear wheels. If backhoe or any rear mounted equipment is installed, remove equipment and left backhoe attaching bracket. Remove the left rear wheel, disconnect left brake and hydraulic linkage and remove bolts retaining left rear fender and side rail to left axle housing. Unbolt left axle housing from center housing and swing axle and housing away from center housing and side frame. Withdraw differential unit from tractor center housing.

When reinstalling axle housing use only one standard gasket to obtain correct preload on differential carrier bearings if gasket is used. Late models use an "O" ring seal. Refer to paragraph 269 for torque values.

OVERHAUL DIFFERENTIAL

272. To overhaul the differential, first remove the unit as in paragraph 265, 269, 270 or 271, and proceed as follows:

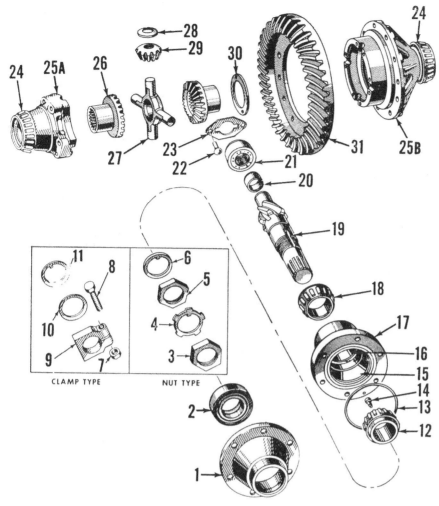

Fig. FO244 — Use jack screws as shown to pull the main drive bevel pinion assembly from wall in center housing.

Fig. FO243 — **Exploded view of main drive gear (ring gear), pinion and differential assembly. On 600, 601 and 2000 Series All-Purpose models, pilot bearings (21) and bearing race (20) are integral and bearing is staked to rear end of pinion gear (19). On all other models, bearing race (20) only is staked to rear end of pinion gear (19), and pilot bearing (21) is retained in center housing by retainer (23) which is riveted to rear side of bearing support in center housing. On Heavy Duty Industrial models, threaded clamp (9) and special washers (10 & 11) are used to retain pinion gear bearings instead of the two nuts (3 & 5), tab washer (4) and keyed washer (6) used on other models.**

1. Seal retainer	10. Washer	18. Rear bearing cone	25A & 25B. Differential
2. Seal	11. Keyed washer	& roller assy.	case
3. Nut	12. Front bearing cone	19. Pinion gear	26. Side gears
4. Tab washer	& roller assy.	20. Bearing race	27. Spider
5. Nut	13. "O" ring	21. Pilot bearing	28. Thrust washers (4)
6. Keyed washer	14. Dowel pin	22. Rivets	29. Pinion gears (4)
7. Nut	15. Front bearing cup	23. Retainer	30. Thrust washers
8. Clamp bolt	16. Rear bearing cup	24. Bearing cone &	31. Rear axle drive
9. Threaded clamp	17. Bearing retainer	roller assy.	(ring) gear

Correlation mark the differential case halves so they can be reassembled in their same relative position. Remove the eight differential case bolts and separate the case halves. See Fig. FO243. Remove spider (27), differential pinions (29) and thrust washers (28). Remove differential side gears (26) and thrust washers (30). Check the differential carrier bearings (24) and renew if required. If bearing cups are to be renewed,

both axle or final drive housings must be off the tractor and the axle shafts removed from their housings to provide clearance for removing the bearing cups.

Differential case and main drive bevel gear are riveted together and may be renewed as an assembled unit, although the bevel ring gear and differential case are available separately. To renew ring gear or differential case if one or the other is still

serviceable, remove ring gear rivets by drilling through rivet heads only with 7/16″ drill and drive or press rivets from gear and case. Special bolts, Part No. NAA-44215-A, and nuts, Part No. 33983-S8, are available for reassembling ring gear to case. Torque the slotted nuts to 90-110 Ft.-Lbs. and install cotter pins. If a high capacity press is available, rivets may be used in lieu of bolts and nuts. On 800, 801, 4000 and all industrial tractors, install rivet heads on differential case side; on all other tractors, install rivet heads in ring gear side of assembly. Upset ends of rivets cold in press. Upset ends must not protrude more than 0.06 above surface. The differential case halves (25A & 25B) are not available separately. Backlash of the bevel gears is not adjustable. Align the matching numbers or correlation marks stamped on differential case halves when assembling. Tighten the bolts to a torque of 90-100 Ft.-Lbs. and lock the bolts with wire if bolt heads are drilled.

MAIN DRIVE BEVEL GEARS
All Models

273. **BEVEL PINION.** To remove main drive bevel pinion, first separate rear axle center housing from transmission as in paragraph 186, 196 or 235 and remove hydraulic power lift cover as per paragraph 312. On Offset type, remove offset housing as outlined in paragraph 267. On 600, 601 and 2000 All-Purpose type, remove differential unit as outlined in paragraph 269. Unscrew the six mounting cap screws and using jack screws as shown in Fig. FO244, remove the pinion and bearing carrier unit. Disassemble pinion and bearings assem-

Fig. FO245 — Adjusting bearing of main drive bevel pinion on models using nut type adjustment.

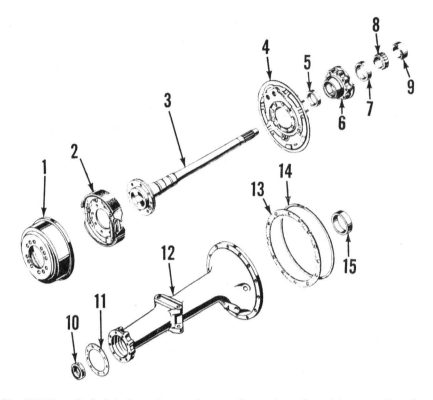

Fig. FO247 — Exploded view of rear axle assembly used on all models except Row Crop and Offset types. Bearing cone and roller (8) is retained in position on axle shaft (3) by lock collar (9) which is installed by heating collar to allow it to be placed on shaft. On cooling, collar shrinks to seize shaft tightly. Gasket (13) is used on models where axle housing (12) is attached to center housing with 7/16-inch studs; "O" ring is used (without gasket) on industrial and other models using 9/16-inch studs to retain axle housing to center housing.

1. Brake drum	5. Grease seal	9. Lock collar	13. Gasket
2. Brake assembly	6. Bearing retainer	10. Oil seal	14. "O" ring
3. Axle shaft	7. Bearing cup	11. Shims	15. Bearing cup
4. Backing plate	8. Bearing cone & roller	12. Axle housing	

Fig. FO246 — Check bevel pinion shaft bearing preload as shown. Scale should read 16-21 pounds.

bly and renew worn or damaged parts.

Reassemble pinion and bearings and adjust bearings with nuts (3 and 5—Fig. FO243) as shown in Fig. FO245, or tighten clamp (9—Fig. FO243) until a torque of 12 to 16 inch pounds is required to turn pinion. When checked as shown in Fig. FO246, the spring scale pull should be 16-21 lbs. After adjustment, bend tabs on lockwasher to secure the shaft nuts or tighten clamp bolt (8—Fig. FO243).

274. BEVEL RING GEAR. To renew the bevel ring gear, remove the differential unit as per paragraph 265, 269, 270 or 271. The procedure for renewal of the ring gear when the differential unit is on the bench is outlined in paragraph 272.

WHEEL AXLE SHAFT, BEARING AND SEAL

All-Purpose, Utility Industrial, Heavy Duty Industrial, LCG and Grove Types

275. BEARING ADJUSTMENT. To adjust the bearings, add or remove shims (11—Fig. FO247) from between one axle shaft bearing retainer and one axle housing until one axle shaft has an end play of 0.002-0.016. To add shims, it will be necessary to remove the axle shaft as follows:

276. R&R AND OVERHAUL. To remove either rear wheel axle shaft, support rear of tractor, disconnect the brake linkage and remove the wheel and tire assembly and brake drum. Unbolt the bearing retainer from axle housing and withdraw the axle shaft, brake and bearing retainer assembly from tractor. The inner oil seal (10—Fig. FO247) can be renewed at this time and must be installed with lip facing the differential.

277. To renew the rear axle shaft, outer grease seal (5), bearing cup (7) or cone and roller assembly (8), it will be necessary to remove the locking collar (9) and press the axle shaft out of the bearing cone and roller assembly.

Use of a special rear axle puller (Nuday tool No. NCA-4235) and a cracking chisel (Nuday tool No. NCA-4235-A1) or equivalent tools are required. The NCA-4235 tool is fitted with drilling guides which can be used to drill a $\frac{9}{32}$-inch pilot hole through the locking collar and to enlarge the hole with a ½-inch drill. The guides align the drill bits so that the drill point will strike the bearing inner race and will not be damaged. The drill motor will speed up when the drill bit reaches the bearing race. Withdraw the drill bit, remove the guide and insert the NCA-4235-A1 cracking chisel with the sharp edges towards the axle shaft. Drive the chisel into the ½-inch drilled hole with a heavy hammer until the locking ring is split. Then, press the axle

1. Bearing retainer
2. Shims
3. "O" ring
4. Bearing cup
5. Bearing cone & roller
6. Pinion & shaft
7. Bearing cone & roller
8. Bearing cup
9. Drop housing
10. Bearing retainer
11. Bearing cup
12. "O" ring
13. Shims
14. Hex nut
15. Washer
16. Bearing cone & roller
17. Spacer
18. Brake drum
19. Axle shaft
20. Oil seal
21. Bearing retainer
22. "O" ring
23. Bearing cup
24. Bearing cone & roller
25. Spacer
26. Bull gear
27. Gasket
28. Oil drain plug
29. Lower cover

Fig. FO248 — Exploded view of Row Crop and Offset rear axle and drop housing assembly. Refer to Fig. FO249 for special components used on Offset right rear wheel axle shaft for implement drive off of rear wheel.

shaft from the bearing. It will aid in removing the axle shaft if the driver at the inner end of the axle shaft is struck with a heavy hammer while turning the puller nut. Note: In case of a broken axle, the two parts can usually be temporarily welded together so that the puller can be used to remove the shaft.

To reassemble, proceed as follows: Renew the bearing cup (7) in bearing retainer (6) if cup is worn or scored. Install new seal (5) with seal lip towards bearing cup and coat lip of seal with Lubriplate. Install seal retainer and brake backing plate (4) on axle shaft. Pack the bearing cone and roller assembly (8) and bearing cup with at least 3 ounces of wheel bearing grease. Use a clean 2¼ inch I.D. pipe, 30 inches long, to drive the bearing cone and roller onto axle shaft. **Be sure** the cone is firmly seated against shoulder on axle shaft.

Heat a new locking collar (9) to 800 degrees F., and quickly place collar over shaft and against the bearing cone. Hold collar in place with the 2¼ inch pipe until collar is cool. CAUTION: If locking collar is overheated, it will not hold the wheel bearing in place. Therefore, use of

a temperature indicating crayon such as "Therm-O-Melt" or "Tempilstick" is recommended. The crayon will melt when rubbed against the collar if collar is heated to proper temperature.

When reassembling, adjust axle end play as outlined in paragraph 275 and tighten retaining nuts to 130-150 Ft.-Lbs.

Row Crop and Offset Types

278. **WHEEL AXLE BEARING ADJUSTMENT.** Factory recommendations are that bearings be adjusted to a certain preload when the bull pinion (final drive) shaft (6—Fig. FO248) and wheel axle shaft oil seal (20) are out of the housing. However, it will not be necessary to remove the oil seal if the job is done as follows: Jack up tractor, drain housing and remove wheel and brake drum. Remove wheel axle shaft inner bearing retainer (10—Fig. FO248 or 10A—Fig. FO249) and identify shims (13—Fig. FO248) for reassembly. Remove the housing lower cover (29). Remove nut (14 or 14A) from inner end of wheel axle shaft, support bull gear (26) and bump wheel axle shaft from housing. Bull gear (26), spacer (25) and outer bearing (24) are re-

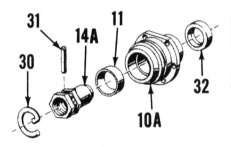

Fig. FO249 — Special components of right wheel axle shaft assembly for Offset models.

10A. Bearing retainer	30. Snap ring
11. Bearing cup	(31 now used)
14A. Drive adapter	31. 3/16-inch pin
	32. Adapter oil seal

moved from bottom of housing. Unbolt and remove the brake backing plate and brake shaft assembly. Remove bull pinion shaft bearing retainer (1) and pull bull pinion shaft (6) from housing.

279. Temporarily install the outer bearing retainer (21) to the final drive housing but do not install the brake backing plate. Using washers under the nuts (instead of the backing plate) to insure full take-up, tighten the retainer nuts to a torque of 130-150 Ft.-Lbs. Install wheel axle

Fig. FO250 — Remove the final drive bull pinion shaft inner oil seal from sleeve by using a suitable puller as shown.

shaft with the bearings and bull gear. Tighten the wheel axle nut (14 or 14A) to 150 Ft.-Lbs. torque. Install the inner wheel axle bearing retainer (10) and tighten the cap screws finger tight only. Attach a spring scale to a cord wrapped around the wheel axle shaft wheel attaching studs and note the pull required to rotate the shaft. This reading is the drag of the oil seal. Now, tighten the inner bearing retaining cap screws to a torque of 60-70 Ft.-Lbs. and again check the spring scale pull required to rotate the wheel axle shaft. Vary the adjusting shims (13) until the scale pull required to rotate the shaft, with the cap screws tightened, is 2 - 5 pounds greater than the first reading taken with cap screws loosened. Adjusting shims (13) are available in thicknesses of 0.003, 0.005 and 0.012.

After the bearings are properly adjusted as outlined above, remove the wheel axle shaft but be careful not to mix or lose the predetermined number of shims (13). Reinstall the final drive (bull pinion) shaft (6), using the original number of shims (2). Reinstall the brake backing plate, wheel axle shaft, bearings and bull gear and tighten the nut on inner end of the wheel axle shaft to a torque of 150 Ft.-Lbs. Install the predetermined number of shims (13) and inner bearing retainer (10).

280. WHEEL AXLE SHAFT, BULL GEAR AND BEARINGS. The procedure for renewing either one of the wheel axle shafts, bull gears and/or bearings is evident when following the procedure for adjusting the wheel axle shaft bearings as outlined in paragraphs 278 and 279.

281. BULL PINIONS, SHAFTS AND BEARINGS. Both of the final

Fig. FO251 — Check preload of Row Crop and Offset type bull pinion shaft with a scale and twine as shown. Refer to paragraph 282.

drive (bull pinion) shafts (6—Fig. FO248) and their bearings can be renewed without removing the final drive housings (9). Removal is accomplished by removing wheel, wheel axle shaft, brake backing plate, bearing retainer (1) and extracting the bull pinion shaft. The bearing cups and cones can now be removed in the conventional manner using suitable pullers. If the bull pinion shaft inner oil seal, shown in Fig. FO250, is to be renewed, it will be necessary to remove the final drive housing and use a puller as shown in Fig. FO250. Before reinstalling the bull pinion shaft refer to paragraph 282.

282. BEARING ADJUSTMENT. Factory recommendation is that the bull pinion shaft bearings be adjusted to a certain preload when the bull gear is out of the housing and bull pinion shaft is out of engagement with the differential side gear. Procedure is as follows: Remove bottom cover (29—Fig. FO248), inner bearing retainer (10), nut, washers and inner bearing cone from wheel axle shaft and bump shaft out of bull gear. Remove final drive housing and mount same in a vise. Install bull pinion shaft and bearing retainer to final drive housing and wrap a length of cord around bull pinion shaft several turns as shown in Fig. FO251. Attach a spring scale to end of cord as shown and measure the pull required to ro-

tate the shaft. Vary the number of shims (2—Fig. FO248) until a scale pull of 7-21 pounds with new bearings or 6-18 pounds with "used" bearings is required to rotate the bull pinion shaft. Shims are available in thicknesses of 0.003, 0.005 and 0.012.

If wheel axle shaft bearings were not previously adjusted, do so as outlined in paragraph 279 **before** reinstalling the bull pinion shaft.

283. AXLE HOUSINGS. To remove or renew right or left hand axle housing interposed between final drive housing (9—Fig. FO248) and center housing, proceed as follows: Jack up tractor and remove wheel and interfering brake linkage. Unbolt final drive housing from outer end of axle housing and lift off final drive unit with a hoist. Remove interfering brake linkage at center housing, then unbolt axle housing from center housing. When removing right hand axle housing lift off the bevel ring gear with the housing, or tie same into position, to prevent it falling out. Housing inner oil seals and bearing cups (in axle housing) and bearing cones of differential can be renewed at this time.

BRAKE SYSTEM

ADJUSTMENT

285. To adjust brakes refer to Fig. FO252 and proceed as follows: Jack up rear wheels, remove adjusting screw cover and turn notched adjuster screw clockwise until wheel no longer can be turned by hand. Then, back-off the adjuster until wheel is free or only a very slight drag is felt. Adjust left brake clevis to equalize the pedals.

R&R SHOES

286. Jack up rear end of tractor and remove the rear wheels. Remove the two screws or spring nuts securing the brake drum to the axle shaft and remove the drum.

Remove the brake shoe hold down cups and springs and using brake pliers, remove the orange adjusting screw spring and the adjusting screw assembly. Remove the anchor pin retaining clips and remove anchor pins from support plate. Remove the two blue return springs (left hand brake has only one) and disconnect the green retracting spring. Lift the shoes from the back plate. Brake shoes are interchangeable from top to bottom.

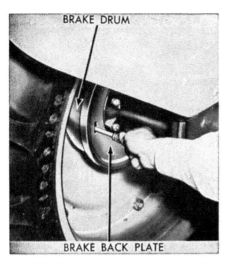

Fig. FO252 — Adjust brake as shown after removing the adjusting screw cover.

Fig. FO253 — Phantom view of PTO external shaft and shift mechanism on four- and five-speed transmission models.

POWER TAKE-OFF

Four speed transmission models and five speed transmission models with single clutches are equipped with a non-continuous type power take-off system which requires engagement of the engine clutch to provide power to the pto output shaft.

Five speed transmission models with dual clutches are equipped with a live or continuous type power take-off system. In this system, a pto countershaft extends through the transmission countershaft where its front end is splined to the pto countershaft drive gear. The pto countershaft gear meshes with the rear end of the pto input shaft to provide constant operation when the engine is running and the pto clutch is engaged.

Both systems are equipped with a manually controlled shifter unit which is attached to rear of transmission and both systems incorporate the output shaft layout shown in Fig. FO253.

"Select-O-Speed" transmission models have two power take-off options: Single-Speed PTO option provides a 540 RPM engine drive PTO speed with an independent PTO clutch. The 2-Speed PTO option provides both 540 and 1000 RPM engine driven PTO speeds and a ground drive PTO speed. Refer to Fig. FO254 for exploded view of output shaft used with either option.

OUTPUT SHAFT

All Four and Five Speed Models and "Select-O-Speed" With Single-Speed PTO

NOTE: Early (1955-56) production tractors were equipped with a 1⅛-inch PTO output shaft. If desired, these early models can

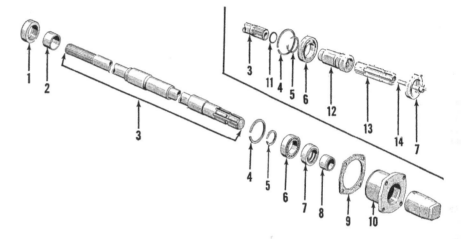

Fig. FO254 — Exploded view of 1⅜-inch pto output shaft used with conventional transmission; output shaft for "Select-O-Speed" with Single-Speed PTO is similar. For 2-Speed PTO in "Select-O-Speed" models, rear end of output shaft (13) can be removed after removing socket head screw (7) to interchange 540 RPM 6-spline shaft and 1000 RPM 21-spline shaft.

1. Oil seal	5. Snap ring	8. Oil seal sleeve	11. "O" ring
2. Bushing	6. Rear bearing	9. Gasket	12. Adapter sleeve
3. Output shaft	7. Oil seal	10. Cover housing	13. Adapter
4. Snap ring			14. Retaining bolt

be converted to 1⅜-inch PTO output shaft by installing the late shaft assembly (Ford part No. 310084) as outlined in paragraph 288. Service of the early 1⅛-inch PTO shaft assembly is similar to that outlined in paragraph 288 except that the oil seal sleeve (8—Fig. FO254) may need to be split before bearing (6) can be pressed from shaft and a new sleeve heated to be installed on shaft.

288. To remove and/or overhaul the pto output shaft, first drain the hydraulic system and the rear axle center housing and proceed as follows: Remove the cap screws retaining the

pto support cover (10—Fig. FO254) to the center housing and withdraw the pto output shaft from the tractor.

Remove the front snap ring (4) from the support cover and press the shaft and bearing out of the cover. After removing the bearing and shaft, the oil seal (7) can be removed from the cover. Oil seal sleeve (8) and bearing (6) can be pressed from rear end of PTO shaft. When installing bearing, be sure it is pressed tightly against snap ring (5) and then press new sleeve (8) tightly against bearing with chamfered end of sleeve to rear.

If the pto shaft has excessive clearance in the front bushing (2), located in the center housing, the bushing and oil seal (1) should be renewed. To renew the oil seal and bushing, it is necessary to detach the rear axle center housing from the transmission. Shaft should have a clearance of 0.0025-0.004 in the bushing.

NOTE: A heavy duty PTO output shaft (Ford part No. 312428) is available for service installation in all four and five-speed transmissions with a 1⅜-inch PTO output shaft, and is interchangeable with the standard duty shaft (Ford part No. 310091). The heavy duty hardened shaft can be identified by an "H" stamped on outer end of shaft.

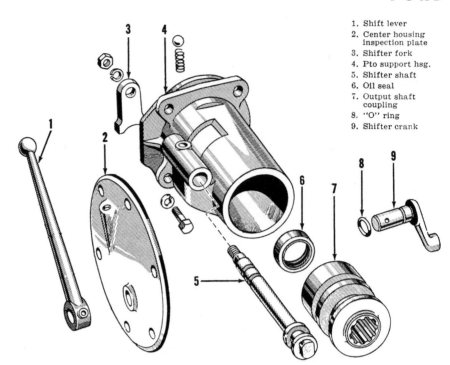

1. Shift lever
2. Center housing inspection plate
3. Shifter fork
4. Pto support hsg.
5. Shifter shaft
6. Oil seal
7. Output shaft coupling
8. "O" ring
9. Shifter crank

Fig. FO255 — PTO support and shifter mechanism used on tractors with conventional four- or five-speed transmission.

"Select-O-Speed" With 2-Speed PTO

289. To remove and/or overhaul the pto output shaft, first drain the hydraulic system and rear axle center housing, remove the cap screws retaining the pto support cover (10—Fig. FO254) to the center housing and withdraw the pto shaft assembly from the tractor.

Clamp the output shaft in a vise, remove socket head cap screw (14) from the end of adapter (13) and withdraw adapter from output shaft and cover assembly. The support cover (10), including adapter sleeve (12), bearing (6) and seal (7) can then be withdrawn from the output shaft. To renew the seal, remove snap ring (4) from front side of support cover and press sleeve and bearing assembly from the cover. Seal can now be renewed. Renew sleeve if the seal surface is worn or if the splines are damaged. Always renew sealing "O" ring (11) when reinstalling the assembly. Center support bushing (2) and seal (1) can be renewed if necessary as outlined in paragraph 288.

SHIFTER UNIT
All Models with Four or Five-Speed Transmissions

290. To remove the shifter unit, first drain lubricant from transmission, hydraulic system and differential and detach the rear axle center housing from the transmission. Unbolt and withdraw the unit from the rear face of the transmission. The overhaul procedure is evident after an examination of the unit and reference to Fig. FO255. Seal (6) which prevents mixing of transmission and hydraulic oil should be renewed if even slightly worn. Use care when installing the support to avoid damaging the seal.

PTO COUNTERSHAFT
All Five-Speed Transmission Models

291. Refer to Fig. FO150. To renew the pto countershaft which is located within the hollow transmission countershaft, remove the pto shifter unit as per paragraph 290 and withdraw the pto countershaft.

PTO COUNTER SHAFT DRIVE GEAR
All Five-Speed Transmission Models

292. Refer to Fig. FO150. To remove the pto countershaft drive gear, it is first necessary to remove the transmission countershaft as outlined in paragraph 204.

PTO INPUT SHAFT
All Five-Speed Transmission Models

293. Refer to Fig. FO150. The pto input shaft is removed in conjunction with removing the transmission input shaft. Refer to paragraph 200.

On five speed transmissions with single clutch, the transmission input shaft and the PTO input shaft are integral.

BELT PULLEY

The belt pulley is supplied as extra equipment and may be mounted and operated in right or left horizontal position or in down vertical position. Do not install pulley in the up vertical position.

NOTE: Two different belt pulley assemblies have been used. The early production belt pulley assembly (shown in Fig. FO256) fits the 1⅛-inch PTO output shaft used in 1955-56 production only. The late production belt pulley assembly is adaptable to the early 1⅛-inch PTO output shaft by using a 1⅛ to 1⅜-inch PTO adapter sleeve. When late production belt pulley assembly is used with current production 1⅜-inch PTO output shaft, a pilot spacer is required between the PTO shaft rear bearing support and the belt pulley housing. Service on both the late and early production assemblies is similar.

R&R AND OVERHAUL

294. Removal of unit requires removal of four cap screws which hold unit to rear axle center housing. To reinstall, engage splines on power take-off shaft and install the four cap screws after locating the unit in desired operating position.

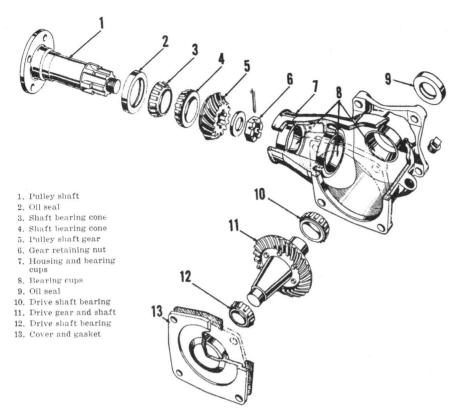

1. Pulley shaft
2. Oil seal
3. Shaft bearing cone
4. Shaft bearing cone
5. Pulley shaft gear
6. Gear retaining nut
7. Housing and bearing cups
8. Bearing cups
9. Oil seal
10. Drive shaft bearing
11. Drive gear and shaft
12. Drive shaft bearing
13. Cover and gasket

Fig. FO256 — Components of belt pulley assembly which is available as optional equipment.

HYDRAULIC SYSTEM

BRIEF DESCRIPTION

Hydraulic lift system incorporates automatic draft control and automatic implement position control. System is basically the same on all models. Fluid supply for the system is contained in the front portion of the center housing where the fluid is isolated from the transmission and bevel ring gear lubricant by oil seals. These seals are located near the front end of the bevel pinion shaft, at the transmission output shafts, and on the steady bushing for the pto shaft.

Elements of the system which include an engine mounted pump, combined main control valve and work cylinder, and a rockshaft are shown in Fig. FO260. The vane type pump shown (early production models only) is provided with an adjustable flow control valve which can be used to vary the raising speed of implements. Recommended fluid for system is a special fluid Ford specfication M4864-A for temperatures above 10° F., or M2C41 for all temperatures. Capacity is eight quarts.

295. Overhaul procedure is as follows: Drain lubricant and remove housing cover (13—Fig. FO256) and gasket. Remove castellated nut (6) from inner end of pulley shaft. This nut must be unscrewed in stages as shaft is being removed. Remove drive shaft and gear (11) out through housing cover opening.

Pulley shaft gear (5) and drive shaft gear (11) are furnished only as a matched set. The mesh and backlash of these gears is fixed and nonadjustable. Install oil seals with lips facing inward.

Adjust pulley shaft bearings to a just perceptible preload by means of pulley nut (6). When only the pulley shaft is in the case, the correct preload is when 5-12 inch pounds (2 - 3 pounds pull on a cord wrapped around 9 inch belt pulley) is required to rotate the shaft in its bearings. Adjust the drive shaft bearings to a slight preload by varying the number of gaskets interposed between cover (13) and housing (7). Correct preload is when 15-34 inch pounds (3 - 5 pounds pull on a cord wrapped around 9 inch pulley) is required to rotate the belt

pulley when the unit is completely assembled. Shim gaskets for drive shaft bearings are available in thicknesses ranging from 0.008-0.023.

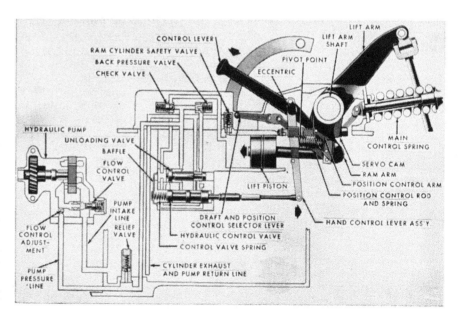

Fig. FO260 — Schematic view of hydraulic system and relative positions of the elements of the system. Early production vane type pump is shown.

TROUBLE-SHOOTING

300. In the Ford hydraulic lift system malfunction of any of its parts will usually show up either in: (a) failure to lift, (b) inability to hold position without slight up or down bobbing motion, (c) erratic action such as over travel or (d) a noisy pump. The probable causes of trouble and methods of checking to locate source of trouble are outlined in paragraphs 301 through 308 which follow.

301. **WILL NOT LIFT.** To determine cause of trouble, first make sure that center housing compartment contains the proper amount and kind of fluid. A drain and fill plug are provided. Insert correct amount of Ford recommended fluid, specification M-4864-A, M2C41 or equivalent to bring fluid level to "FULL" mark on dip stick.

302. SELECTOR VALVE SETTING. Place selector lever (23—Fig. FO263) in the down or Constant Draft Control position and touch control lever (34) in upper or raise position. If system fails to lift, move the selector (23) lever to horizontal position and check the following: If system lifts when selector lever is in the horizontal position but does not lift when same lever is in down position, proceed as follows: Remove PTO shifter plate from left side of center housing and move touch control lever (34) to stop at top of quadrant. Move selector lever (23) to the down position. Using the Constant Draft end of gage block (Nuday tool No. NCA-502) check the gap between control valve land and machined surface of valve housing as shown in Fig. FO-261. If gage cannot be entered or fits the gap loosely, adjust the turnbuckle as shown until the gage fits the gap snugly. If system still fails to lift, check the pump prime as follows:

303. PUMP PRIME. To check the pump prime on models equipped with a vane type pump, remove the pipe plug from the lower right front side of the center housing and place a container under the opening. With the touch control lever down and engine running, oil should flow from the pipe plug hole. If there is no oil flow, prime the pump by applying air pressure to the center housing via the hydraulic system dip stick hole or by racing the engine. If flow is not obtained after priming, it will be necessary to renew or recondition the pump. If oil flow is obtained (either before or after priming) but system will not lift, refer to paragraph 304.

To check the pump prime on models equipped with a piston type pump, remove the Allen head plug in pump cover, crank the engine with the starting motor and check for oil flow from the pipe plug hole. If oil does flow from the pipe plug hole, reinstall the plug while the oil is still flowing to prevent the entry of air. If oil does not flow, attach a hose to bleed hole, insert free end of hose in hydraulic filler opening, and start engine. When a full, steady oil flow is obtained, stop engine, remove hose and reinstall bleed plug. If flow cannot be obtained, renew or recondition the pump. If oil flow is obtained (either before or after priming), but system will not lift, refer to paragraph 304.

304. PUMP OUTPUT PRESSURE. On early model tractors equipped with a vane type pump, remove the flow control valve spring retainer plug from front end of pump flow control valve body and withdraw the spring and flow control valve. (See Fig. FO273). Thoroughly clean the valve and remove any burrs from same with crocus cloth. Reinstall the valve, making certain that it moves freely in its bore. Reinstall the spring and plug, using a new "O" ring.

On all models, remove the pipe plug (92—Fig. FO269) from lower right front side of center housing and connect a high reading (at least 2500 psi) gage to hole from which the plug was removed. Remove the accessory plate (11—Fig. FO263) from the lift cover. Install the accessory plate on the lift cover with the section marked (B) in Fig. FO270 over the pressure tube using a paper gasket between plate and lift cover. Start engine and run at 600 rpm; at which time, the relief valve should unseat as indicater by buzzing or bubbling of oil in center housing near the valve and the pressure gage should register 1950-2050 psi as the relief valve unseats. If the relief valve unseats at less than 1950 psi, renew the valve and recheck. If valve doesn't unseat and the pressure remains low, renew or recondition the pump.

305. VALVE OPERATION. If system still does not lift and the system operating pressure is known to be O. K., check for unloading valve (45—Fig. FO264) stuck in bypass position and/or for control valve not going into lift position.

306. **BOBBING.** (HICCUPS). To determine the cause of leakage, mount a heavy implement and raise the three-point linkage. Remove the hydraulic filler cap and visually check rear end of cylinder for leakage around piston. If a leak is noted; renew the piston seal.

If piston was not leaking, shut off engine with implement raised. If implement falls about 6 - 12 inches, then stops; or if rate of fall decreases noticeably after falling about one foot, renew the control valve and bushing as outlined in paragraph 315.

If implement falls all the way to ground at a steady rate, a leaking check valve or safety valve is usually indicated. Remove the accessory plate and renew the safety valve (42—Fig. FO264); then repeat the test. If leak still occurs, renew the check valve and seat as outlined in paragraph 313.

Additional points to check are the "O" rings located between lift cylinder and cover and the leather back-up washer (56—Fig. FO264) and "O" ring (55) on the lift cylinder piston. Renew lift cylinder (35) if cylinder bore is scored or otherwise damaged. While lift cover is off, make sure that valve linkage operates without binding.

307. **ERRATIC ACTION.** This trouble is manifested by erratic response to movement of the touch control lever resulting on over-travel or over-correction. Usual cause of trouble is a sticking control valve, binding in the control valve linkage, or sticking unloading valve. Remove lift cover, inspect for free movement and free up if sticky. At same time, make sure that draft control link swivel (70—Fig. FO267) slides freely on bushing (72). Correct any misalignment and renew any worn parts.

Another probable cause, which can be checked with lift cover installed, is binding of top link rocker arm on pivot pin at rear of differential center housing, especially on Row Crop and Offset models.

308. PUMP LOSES PRIME. Loss of prime is usually caused by a leak in the suction (intake) side of the system. Also, loss of prime may be caused by low hydraulic fluid level or use of improper oil in system which causes excessive foaming condition.

Check fluid level with lift arms raised and with all remote hydraulic cylinders extended; then, add sufficient M-4864-A or M-2C-41 hydraulic fluid to bring fluid level to full mark on dipstick located at left side of center housing. If oil appears excessively foamy after operating system for a period of time, drain system and refill with correct type of hydraulic fluid.

Renew pump drive shaft seal and, if sealing surface on shaft is grooved, renew shaft. Refer to paragraph 326 for vane type pump and to paragraph 327 for piston type pump. On early piston type pump with socket head plugs in pump body, remove the plugs and coat threads with white lead. Re-install plugs securely.

If loss of prime persists, renew the sealing "O" rings between hydraulic pump and manifold and between manifold and transmission. Check the pump manifold for cracks and be sure that mounting surfaces are flat. Re-new the sealing "O" rings between center housing and transmission.

SYSTEM ADJUSTMENTS

309. MAIN CONTROL SPRING. Adjust tension on the main control spring (2—Fig. FO263) by removing the rocker arm pin from yoke (1) and turning the yoke in or out. For average conditions, adjustment of main control spring is correct when spring can just be rotated by thumb and fingers of one hand. Under heavy draft conditions, it may help to turn the yoke one-half turn tighter than this adjustment; or under light draft loads, loosen yoke one-half turn from this adjustment.

Before making constant draft adjustment as outlined in paragraph 310, or position control adjustment as outlined in paragraph 311, be sure that main control spring is adjusted so that there is no end play of spring or plunger (8).

310. CONSTANT DRAFT. Although this adjustment can be performed with lift cover installed, it is recommended that the lift cover be removed and the position control be adjusted also as outlined in paragraph 311. Any change in adjustment of constant draft linkage also affects the adjustment of position control linkage. Always make constant draft adjustment first as follows:

With cover assembly mounted as shown in Fig. FO261, attach lift arm locating gage, Nuday tool No. NCA503 or NCA-503-B, to lift cover and insert locating pin through lift arm and gage. Note: Be sure all gasket material is scraped from lift cover before attaching gage.

Place selector lever (23—Fig. FO-263) in the draft control (down) position and move the touch control lever (34) against the stop at top of quadrant. At this time the gap between the shoulder on the control valve spool and the machined surface

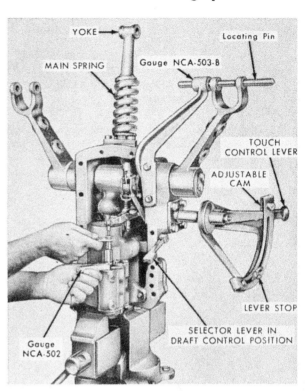

Fig. FO261—View showing method of adjusting draft control linkage using Nuday gage NCA-502 and lift arm positioning gage, Nuday tool No. NCA-503-B. Refer to paragraph 310 for procedure. Refer to Fig. FO262 for view showing method of adjusting position control linkage.

of the lift cylinder casting should just admit the "draft" end of thickness gage, Nuday tool No. NCA-502, as shown in Fig. FO261. If gage will not enter the gap, or the gap is wider than the thickness gage, loosen the locknut (51—Fig. FO264) on link (50) and adjust the turnbuckle (52) until draft end of gage is a sliding fit in the gap. Tighten the locknut and re-check the adjustment before proceeding further.

If special Ford gages are not available, adjust the constant draft linkage as follows:

Mount the lift cover assembly in a vise as shown in Fig. FO261, adjust main control spring as outlined in paragraph 309 and move the lift arms to their extreme lower position until the end of the ram piston contacts the end of the cylinder. Move the lift arms back from the extreme lower position approximately ½ inch when measured at the pin hole in the lift arm. Lock the lift shaft in this position by tightening the retaining cap screws on the ends of the lift shaft on late production models, or by removing washer (83—Fig. FO268) and installing spacer between lift arm (82) and washer (83) on early production models, so that the position of the shaft will be maintained. Move the selector (small) lever to the draft

control (down) position and the touch control lever against the cam stop in the raised position as outlined for gage adjustment above.

Measure gap between machined surface of lift cylinder casting and exposed land on control valve and adjust to 0.395-0.397 by means of the control valve turnbuckle. Leave the lift shaft positioned as above, and proceed with implement position adjustment as outlined in paragraph 311.

Note: If both constant draft and implement position adjustments cannot be obtained because of lack of adjusting threads, remove draft control fork (73—Fig. FO267) and check it against a new one. Because of its shape, a bent control fork is very hard to detect without this comparison. Also, check all linkage for excessive wear.

311. POSITION CONTROL. To make this adjustment, first adjust the main control spring as per paragraph 309 and the constant draft control as per paragraph 310. With cover mounted in vise as shown in Fig. FO-262, place selector lever in position control (parallel to cover) as shown and touch control lever against stop at forward or down end of quadrant as shown.

At this time, the gap between control valve land and the machined surface of lift cylinder should just ad-

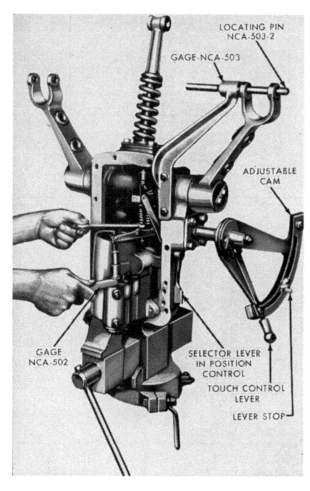

LOCATING PIN
NCA-503-2

GAGE-NCA-503

ADJUSTABLE CAM

GAGE NCA-502

SELECTOR LEVER
IN POSITION
CONTROL

TOUCH CONTROL
LEVER

LEVER STOP

Fig. FO262—View showing method of adjusting position control linkage. Always make draft control linkage adjustment before checking adjustment of position control linkage. Refer to paragraphs 310 and 311 for procedure and to Fig. FO261 for view showing adjustment of draft control linkage.

mit the "position" end of special thickness gage, Nuday tool No. NCA-502. If gage will not enter or fits loosely, loosen the locknut on position control rod (62—Fig. FO267) and turn the hexagon head of the position control rod while holding stamped cam plate (67) with a suitable wrench until "position" end of gage fits snugly as shown. Note: Failure to hold stamped plate (67) may result in a sheared locating pin (68) on position control arm (65).

If special Ford gages are not available, adjust the constant draft linkage as outlined in paragraph 310 and proceed as follows:

Move selector (small) lever into the implement position (parallel with cover flange) and touch control lever into the lower position against the lower stop as shown in Fig. FO262.

Measure clearance between machined surface of lift cylinder casting and exposed land of control valve and adjust to 0.448-0.450 by adjusting the free length of position control rod as outlined for gage adjustment.

LIFT COVER AND CYLINDER, OVERHAUL

The lift cover assembly includes the rock (lift) shaft, control quadrant, lift (work) cylinder, main control spool valve, unloading valve, safety valve, check valve and back pressure valve. The pump relief valve is located in the bottom of the center housing.

312. R&R LIFT COVER AND CYLINDER ASSEMBLY. Move touch control lever to bottom of quadrant, selector lever to draft control position, and step on lift links to exhaust all oil from the hydraulic cylinder. Remove the pin from top link rocker arm and yoke (1—Fig. FO263) and remove pins from lift links and lift arms (83—Fig. FO268). On Offset models, unbolt the accessory plate manifold from offset lift cover adapter casting. Remove tractor seat and remove cap screws that retain lift cover to center housing or offset adapter casting. Break the gasket seal loose and remove lift cover assembly from tractor.

Always use new gasket and two new "O" rings when reinstalling lift cover. Do not apply any gasket sealer or grease to gasket surfaces. Tighten the lift cover retaining cap screws to a torque of 45-65 Ft.-Lbs.

313. CHECK VALVE AND BACK PRESSURE VALVE. Except on Offset models, it is necessary to first remove the lift cover and cylinder assembly as outlined in paragraph 312 to permit removal of the check valve and back pressure valve.

Refer to Fig. FO263. Using a six-point box end wrench or socket, unscrew the hex head plug (13). Using needle nose pliers, remove the pilot (14) and "O" ring. Then, withdraw the spring (16), guide (17) and check valve (steel ball) (18).

The check valve seat (20) is threaded on inside diameter. Screw the special puller (Nuday tool No. NCA-997) into valve seat until finger tight, then back off one turn. While holding the threaded shaft from turning, pull the valve seat by turning the hex nut on puller down against spacer. Note: It is possible that the check valve seat may be broken, either during installation or removal, and only the front half will be removed with the special puller. If this occurs, proceed as follows: Remove the accessory plate and lift cylinder from lift cover. Grind down a ¼-inch steel nut until it can be inserted between the rear half of the broken seat and the back pressure valve (21) through an oil passage hole and pull remaining part of seat with a ¼-inch bolt.

Inspect front end of check valve seat (20); if seat for check valve ball is nicked or chamfered, renew the check valve seat. Seating surface for back pressure valve on rear end of check valve seat is not critical. Renew the springs (16 and 22) if rusted, cracked or worn. Always renew the "O" rings (15 and 19).

NOTE: At tractor Serial No. 46496 of 1963 and later production series, the check valve seat was changed from a sharp edge to a chamfer type seat and the check valve ball diameter was changed from 7/16-inch to 15/32-inch. The previously used seat cannot be used with the larger diameter check valve ball and vice versa. However, the seats are interchangeable if the correct size check valve ball is used. The previous type sharp edge seat will no longer be available when service stocks are exhausted. Affected part numbers are as follows:

	Old Part Number	New Part Number
Valve Seat	NCA-997-A	C3NN-997-A
Valve Ball	354069-S	351666-S

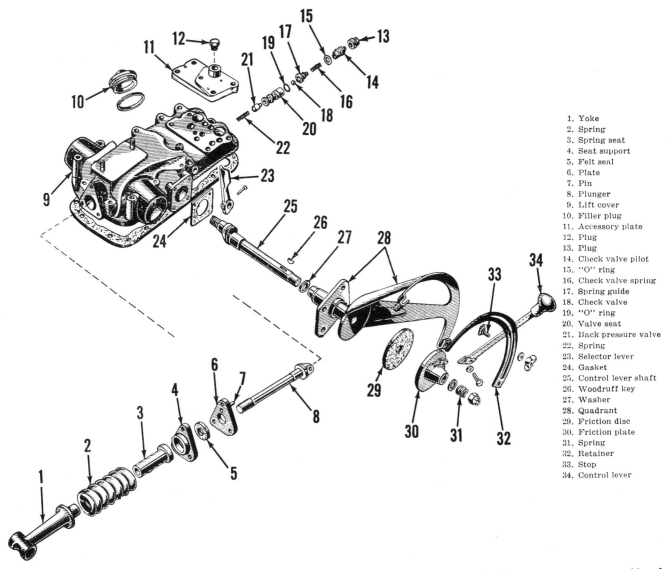

Fig. FO263 — Exploded view of hydraulic lift cover and related parts. Lift cylinder (See Fig. FO264) attaches to bottom side of cover. Linkage shown in Fig. FO267 connects control valve to draft control plunger (8), touch control lever shaft (25) and to position control cam on lift ram arm (80—Fig. FO268). Note that selector lever (23) is shown in "up" position, but should be in forward or down position.

1. Yoke
2. Spring
3. Spring seat
4. Seat support
5. Felt seal
6. Plate
7. Pin
8. Plunger
9. Lift cover
10. Filler plug
11. Accessory plate
12. Plug
13. Plug
14. Check valve pilot
15. "O" ring
16. Check valve spring
17. Spring guide
18. Check valve
19. "O" ring
20. Valve seat
21. Back pressure valve
22. Spring
23. Selector lever
24. Gasket
25. Control lever shaft
26. Woodruff key
27. Washer
28. Quadrant
29. Friction disc
30. Friction plate
31. Spring
32. Retainer
33. Stop
34. Control lever

Insert spring (22) and back pressure valve (21), open end first, into lift cover. Be sure that valve is in small diameter (rear) part of bore in lift cover and that the valve works freely. If lift cover is in place on Offset models, remove the accessory plate (11) and hold the back pressure valve in place with screwdriver inserted through oil passage hole while check valve seat is being installed. On other models, clamp lift cover solidly in a vise with front end of lift cover up.

Remove spacer from valve seat removal tool (Nuday tool No. NCA-997) and back hex nut off as far as possible. Thread tool tightly into check valve seat with a flat washer between seat and shoulder on tool. (Washer

O.D. must be as small or smaller than valve seat diameter.) Install new "O" ring (19) on check valve seat, lubricate valve seat and "O" ring and, using a steel hammer, drive valve seat into lift cover until it bottoms on shoulder in bore. Note: Most mechanics prefer to remove accessory plate so that back pressure valve can be held in proper position while driving valve seat into lift cover.

Insert check valve ball (18), spring guide (17) and spring (16). Install new "O" ring (15) on check valve pilot (14), lubricate "O" ring and push pilot into bore. Install and tighten retainer plug (13) to a torque of 45-55 Ft. Lbs. Note: Over-tightening plug will make it very difficult to remove and will not improve seal-

ing. If accessory plate has been removed, install using new "O" rings and tighten the $\frac{5}{16}$-inch center cap screw to 12-15 Ft. Lbs. Tighten the $\frac{7}{16}$-inch cap screws to 45-65 Ft.-Lbs.

314. R&R LIFT CYLINDER ASSEMBLY. To remove the lift cylinder assembly (See Fig. FO264), first remove the lift cover assembly from tractor as outlined in paragraph 312. Disconnect link (50—Fig. FO264) from control lever (76—Fig. FO267) and remove link and turnbuckle assembly from control valve. Remove the accessory plate (11—Fig. FO263); then, unbolt and remove cylinder assembly from lift cover.

Use new "O" rings when assembling cylinder and lift cover. Tighten the cylinder retaining cap screws to a torque of 60-80 Ft.-Lbs.

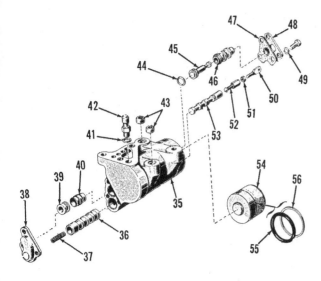

Fig. FO264 — Exploded view of hydraulic lift cylinder.

35. Cylinder
36. Control valve bushing
37. Spring
38. Baffle plate
39. Unloading valve plug
40. Unloading valve bushing (Large I.D.)
41. Copper gasket
42. Safety valve
43. Hollow dowels
44. "O" ring
45. Unloading valve
46. Unloading valve bushing (Small I.D.)
47. Gasket
48. Retainer plate
49. Soft washer
50. Link
51. Locknut
52. Turnbuckle
53. Control valve
54. Lift piston
55. "O" ring
56. Back-up ring

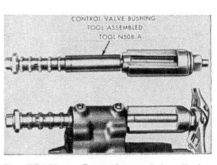

Fig. FO265 — Extraction and installation of control valve bushing is accomplished by using Nuday tool N508-A as shown.

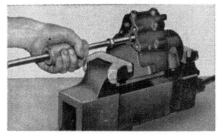

Fig. FO266—Method of removing pressed-in unloading valve plug.

315. SAFETY VALVE. The system safety valve (42—Fig. FO264) is threaded into the top of the lift cylinder casting and functions to prevent shock loads from damaging the system.

As the pressure setting of the safety valve is above the normal operating pressure of the system, it cannot be checked unless a pressure test pump, a high pressure gage (minimum of 3000 psi) and suitable connections are available. If condition of safety valve is questionable and test equipment is not available, renew the valve. Note: **Do not** attempt to reset safety valve operating pressure.

NOTE: The safety valve can be removed with lift cover installed on tractor after removing the accessory plate (11—Fig. FO-263). Do not attempt to remove the valve with screwdriver; use a deep wall socket.

316. CONTROL VALVE, UNLOADING VALVE AND BUSHINGS. To service the control valve, unloading valve and bushings, first remove the lift cylinder assembly as outlined in paragraph 314. Then, unbolt and remove baffle plate (38—Fig. FO264) and withdraw control valve spring (37). Unbolt and remove retainer plate (48) and withdraw control valve spool (53). Refer to paragraph 317 for further information on servicing control valve and bushing and to paragraph 318 for servicing unloading valve and bushings.

When reinstalling retainer plate (48), use new gasket (47) and **be sure** that sealing washer (49) is installed under cap screw located between control valve spool and cylinder (lift piston) bore. Renew washer (49) if damaged in any way.

317. CONTROL VALVE AND BUSHING. Inspect the lands on control valve spool and renew valve and bushing if erosion or scoring is visible, or if trouble shooting checks indicated leakage of oil at control valve. It is recommended that neither the control valve or the bushing be renewed without renewing the mating part.

After removing the control valve spool as outlined in paragraph 316, the control valve bushing can be removed using a special puller (Nuday tool No. N-508-A) or can be removed in a press by using suitable bushing drivers. Be careful not to score bushing bore in cylinder casting with oversize or rough bushing drivers.

The control valve bushing is available in two different outside diameter size ranges which are color coded blue (smallest diameter) and yellow (largest diameter). The lift cylinder casting is also color coded near the control valve bushing bore to indicate the size range bushing required. Always renew the bushing with new bushing having same color code. Note: It is possible that some cylinders may have a white color code; in that case, install a blue color code service bushing.

When installing new bushing, be sure that bore in lift cylinder is clean and free of nicks or burrs. Lubricate both the bushing and bushing bore and insert end of bushing having widest land into the bore at open end of cylinder. Press the bushing into place with special puller or in a press so that ends of bushing are flush with machined surfaces of lift cylinder casting.

Control valve spools are available in three different size ranges which are color coded white (smallest spool diameter), blue, and yellow (largest spool diameter). The correct size control valve spool can be determined only by selective fit **after** the bushing has been pressed into the cylinder casting. Lubricate the valve spool and bushing and insert spool with open end towards open end of cylinder. A drag should be felt on the valve when moving it in its normal range of travel. If valve binds, select a smaller diameter valve spool. If valve moves freely through bushing, select a larger diameter valve spool. Note: As the color code indicates a size range only, a valve spool of one color code may fit correctly while other valves having same color code may fit too tight or too loose. After reinstalling valve, retaining plate, spring and baffle plate, check to see that spring returns control valve quickly when valve is depressed and released; if not, valve is either fit too tightly or dirt or other foreign material is causing valve to bind.

318. UNLOADING VALVE, PLUG AND BUSHINGS. After the control valve spool has been removed as outlined in paragraph 316, the unloading valve can be removed as follows: Thread an impact puller adapter into the unloading valve plug (See Fig. FO266) and pull plug from cylinder casting. With a thin rod inserted through end of bushing (46), push unloading valve (45) and "O" ring (44) from cylinder.

Inspect bores of bushings (40 and 46) and renew bushings if scored, excessively worn or other defects are noted. To remove bushings, use special puller (Nuday tool No. N-508-A) or press and suitable bushing drivers. Pull or press bushings out towards closed end of cylinder. Bushings are available in two outside diameter size ranges which are color coded blue (smallest diameter) and yellow (largest diameter). Also, cylinder casting is color coded near unloading valve bushing bore. Always renew the bushings with new bushings having same color code. Note: It is possible that some cylinders may have a white color code; in that case, install blue color coded service bushings.

When installing new bushings, be sure that bore in lift cylinder is clean and free of nicks or burrs. Lubricate the bushings and bushing bore. The front (large I.D.) bushing (40) has notches in each end; insert this bushing in bore at open end of cylinder with end having large notch out (towards open end of cylinder). Place end of rear (small I.D.) bushing (46) against the large notch end of front bushing and press both bushings into cylinder until land on rear bushing (46) is flush with machined surface of cylinder at rear (open) end of cylinder.

Renew the unloading valve (45) if scored, excessively worn or otherwise damaged. Valve is available in one size only. Lubricate the valve and insert it in the bushings without the "O" ring (44). Valve should be a free sliding fit in the bushings. Remove valve and install "O" ring (44), lubricate valve and "O" ring and reinstall in bushings. The "O" ring should impart a slight drag when moving the valve back and forth in the bushings. If valve binds or sticks, or moves freely as without the "O" ring being installed, the valve should be removed and another "O" ring installed; there may be a slight difference in "O" ring size. CAUTION: Never install an "O" ring of unknown quality in this location. Some "O" ring materials shrink or swell when subjected to hydraulic fluid and heat and there-

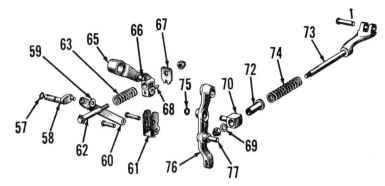

Fig. FO267 — Exploded view of control linkage for Ford hydraulic system.

57. "O" ring	62. Lift position rod	68. Pin	74. Spring
58. Arm	63. Spring	69. Washer	75. Snap ring
59. Washer	65. Position control arm	70. Swivel	76. Control valve lever
60. Link	66. Pin	72. Bushing	77. Pin
61. Cam	67. Cam plate	73. Draft control fork	

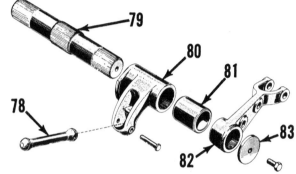

Fig. FO268 — Exploded view of lift ram arms and shaft as removed from lift cover shown in Fig. FO261.

78. Connecting rod
79. Lift arm shaft
80. Lift ram arm
81. Bushing
82. Lift shaft arm (L.H.)
83. Washer

by cause malfunction of the hydraulic system.

Reinstall the unloading valve plug with threaded hole facing out. If for some reason the plug is being renewed, select a new plug having the same color code as on the cylinder casting near the unloading valve bushing bore. New plugs are available in three different size diameters and are color coded white, blue and yellow.

319. LIFT CYLINDER PISTON AND SEALING RINGS. After removing the lift cylinder as outlined in paragraph 314, inject air pressure into the cylinder pressure port to eject piston from cylinder. (Cover safety valve hole with thumb if valve has been removed.) Care should be taken when removing the piston as it may be ejected with considerable force depending upon amount of air pressure.

Inspect the cylinder bore and renew cylinder if cylinder wall is deeply scored or cracked. Minor imperfections can be removed with a hone. New service cylinder is fitted with control and unloading valve bushings (36, 40 & 46—Fig. FO264).

Renew the piston if excessively worn or scored. Do not attempt to remove the black coating from new service piston; the coating is applied to help prevent piston scoring.

Soak new leather back-up ring (56) for a few minutes in hot oil or water to prevent it from breaking when being stretched to install on piston. Install back-up ring with smooth side of leather towards open end of piston. The ring can be placed over a piston ring compressor, the compressor expanded to fit over the piston skirt and the ring pushed onto the piston as an aid in installation. Install the "O" ring (55) in front of leather ring so that "O" ring mates against rough side of leather ring.

320. LIFT SHAFT, ARMS AND BUSHINGS. The lift shaft (79—Fig. FO268), bushings (81) and lift cylinder arm (80) can be removed after removing the lift cover assembly as outlined in paragraph 312.

Early production lift shaft length was slightly greater than the combined width of the lift cover and lift arms, and the lift arms were retained by two cap screws in each end of the

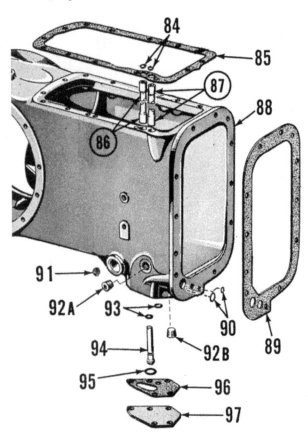

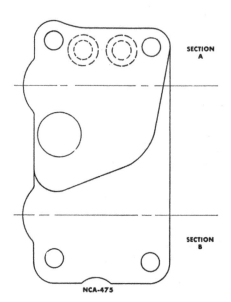

Fig. FO269—View showing location of hydraulic system relief valve (94), pressure tube (87) and return tube (86).

84. "O" rings
85. Cover gasket
86. Return tube
87. Pressure tube
88. Center housing
89. Gasket
90. "O" rings
91. Drain plug
92. Pressure passage plugs
93. "O" rings
94. Relief valve
95. "O" ring
96. Gasket
97. Cover plate

Fig. FO270—Accessory plate showing test set-ups. Note sections A and B

shaft. To remove shaft, bend locking tabs at one end of the shaft down, remove the retaining cap screws, washer (83) and lift arm. The shaft can then be withdrawn from the opposite side of lift cover. Excessive end play of early production lift shaft assembly can be corrected by installing new lift arms, washers (83) and/or lift cover casting. Note: Machine washers, available locally, can be fitted over lift shaft between lift arms (82) and washer (83) to reduce end play also. Be sure that lift arms are free to fall of their own weight.

Late production lift shaft length is less than combined width of lift cover and lift arms, and end play is controlled by tightening self-locking retaining cap screws against the lift arm retaining washers (83). Tighten the retaining cap screws equally at each end of shaft so that there is no end play in the lift shaft assembly and the lift arms will just fall of their own weight. Lift shaft can be removed after removing cap screw and retaining washer from either end of shaft. Late production lift shaft (79) can be installed on early production models by obtaining and installing spring steel washers with single bolt hole (83) and self-locking retaining cap screws.

Lift shaft (79), cylinder lift arm (80) and lift arms (82) have master splines. Bushings (81) are a hand push fit in the lift cover.

321. CONTROL LINKAGE. To completely service the control linkage, remove the lift cylinder as outlined in paragraph 314 and remove the lift shaft and lift arms as outlined in paragraph 320. Refer to Figs. FO263 and FO267; then, proceed as follows:

Unscrew the yoke (1) and remove the main control spring (2) and spring seat (3). Remove the three retaining cap screws and the retainer (4), felt (5) and plate (6). Note: In some cases, the yoke (1) may be seized to the plunger (8). If so, do not apply excessive torque to remove the yoke or the control linkage may be seriously damaged. Heat the yoke with a torch, or cut plunger (8) in two with hacksaw between coils of spring and renew yoke and plunger. Threads of yoke and plunger should be liberally lubricated with heavy grease on reassembly.

Remove snap ring (75) and disengage swivel from control lever (76). Remove fork (73) with spring (74), bushing (72), swivel (70) and plunger (73) attached. Disassembly procedure for this unit is evident. Check the

fork (73) for any bends in rod portion and for looseness of pin retaining fork to plunger; renew fork, pin and/or plunger if defects are noted.

Remove nut from inner end of control lever shaft (25) and remove washer and control lever (76) from shaft. Remove nut from outer end of shaft and remove spring (31) and flat washer. Unbolt and remove touch control lever (34) from friction plate (30). Remove friction plate, Woodruff key (26) and friction disc (29) from shaft. Note: In some cases, friction plate (30) may be seized to shaft; if so, split the friction plate with a sharp chisel. Unbolt quadrant (28) from lift cover and pull quadrant from control lever shaft. Remove cotter pin and disconnect link (60) from selector lever arm (58). Withdraw control lever shaft and position control arm (65) assembly from lift cover; procedure for disassembling this unit is evident. Drive roll pin from selector lever (23), remove lever from arm (58) and remove arm from lift cover.

Renew any linkage that is bent or where pivot holes are worn. Renew plate (6) if pin (7) is loose or worn or if head of plunger (8) has seated into plate. Pins (66, 68 and 77) in position control arm (65) and control lever (76) are renewable if arm and/-or lever are otherwise serviceable.

Reverse disassembly procedure to reassemble and reinstall linkage in

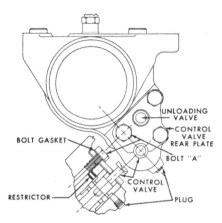

Fig. FO271 — View showing installation of special restrictor in hydraulic circuit. Refer to text.

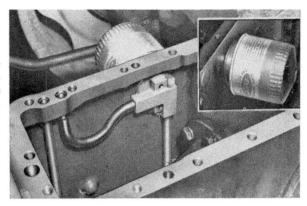

Fig. FO272 — View showing installation of hydraulic oil filter that is available for field installation on all models with hydraulic lift system.

lift cover. Tighten the retaining nut on front end of fork (73) until washer (69) is tight against shoulder on shaft part of fork. Be sure to apply grease to the threads on yoke (1) and plunger (8) and also grease outer end of control lever shaft (26) and bore of friction plate (30). After unit is completely reassembled and lift cover is reinstalled, adjust tension of friction plate spring (31) so that touch control lever will stay in desired position.

NOTE: Before reinstalling lift cover, be sure all linkage moves freely and adjust draft and position control linkage as outlined in paragraphs 310 and 311.

SPECIAL HYDRAULIC VALVING AND HYDRAULIC OIL FILTER KIT

Parts are available through Ford Tractor Dealers to modify the Ford Hydraulic System for increased sensitivity to changes in draft. A new type control valve spool, unloading valve front bushing and a restrictor orifice provide slower and smoother draft corrections and a faster rate of implement drop. This is especially desirable with light draft implements where close control of depth is required. Refer to following paragraphs.

322. RESTRICTOR ORIFICE. When installing the special control valve and unloading valve bushing (See paragraph 323), a restrictor orifice, Ford part No. NCA-A951A, must be installed as follows: Refer to Fig. FO271. After removing the control valve bushing as outlined in paragraph 317, remove the socket head plug from lift cylinder casting as indicated in Fig. FO271. Using a rod

or punch, press the restrictor into the drilled passage between the control valve bushing bore and the threaded hole into which the cap screw (BOLT "A") is threaded. Be sure that the restrictor does not extend into either the threaded hole or bushing bore.

323. SPECIAL CONTROL VALVE AND UNLOADING VALVE BUSHING. After installing restrictor as outlined in paragraph 322, renew the control valve bushing (using regular type bushing) and the control valve spool (using new type valve spool) as outlined in paragraph 317. Remove the unloading valve plug, valve, and bushings as outlined in paragraph 318. Renew the "O" ring on unloading valve and renew the unloading valve and/or rear bushing if required. Install the new type front bushing, rear bushing, valve, "O" ring and plug following procedures outlined in paragraph 318. Carefully readjust the hydraulic linkage as outlined in paragraphs 310 and 311.

324. HYDRAULIC OIL FILTER KIT. Refer to Fig. FO272. The oil filter kit may be installed on all models having a hydraulic lift system, and installation of kit is especially recommended when installing the special control valve and unloading valve bushing. Complete installation instructions are provided with the oil filter kit. After installing kit, use regular engine oil filter (throw-away type, Ford part No. B8NN-6714-A).

PUMP, MANIFOLD, RELIEF VALVE AND OIL TUBES

Early model tractors were fitted with a Vickers vane type pump which has a capacity of 4.7 gpm at 1650 engine rpm. Beginning with serial number 66849 of 1955-57 production, all tractors were equipped with a piston type pump which has capacity of 4 gpm at 2000 rpm. Pump relief valve (94—Fig. FO269) is located in the bottom of the center (differential) housing under the small sump cover (97). The hydraulic system pres-

sure and intake tubes extend through the transmission to connect the hydraulic manifold at the front to the sump in the center housing.

325. R&R PUMP. Removal procedure is as follows: Disconnect the Proof-Meter cable. Remove the pump-to-manifold nuts and the pump-to-engine cap screws. After lifting off the pump, discard the gasket and "O" rings and close the manifold opening with a clean cloth.

When reinstalling the pump, tighten the two pump-to-engine bolts securely before tightening the pump-to-manifold nuts. Remove the proof-meter drive unit and check to see that there is some backlash between the pump drive gear and camshaft gear. If no noticeable backlash is present, remove the pump and install a second gasket. If no backlash can be detected when using two pump-to-engine gaskets, use gasket to cut metal shims as required. Pump will be noisy if there is no backlash between pump and camshaft gear.

326. OVERHAUL VANE PUMP. NOTE: Vane pump may be replaced with piston pump and manifold. Rather than make extensive repairs on vane pump, it is recommended that piston pump and manifold be installed.

Recommended order of disassembly is as follows: Refer to Fig. FO273. Remove flow control valve body. Carefully separate the gear housing from pump and shaft by removing Proof-Meter drive adaptor slotted gear retaining bolt and the two long bolts retaining the pump cover to gear housing. Separate the cover and cam ring from pump body by removing key and spacer from pump shaft and the two short bolts at cover end of pump. Shaft and bearing can be pressed rearward out of body. A slide hammer jaw puller may be used to extract the shaft seal from body. Remove plug for access to flow control valve plunger and spring.

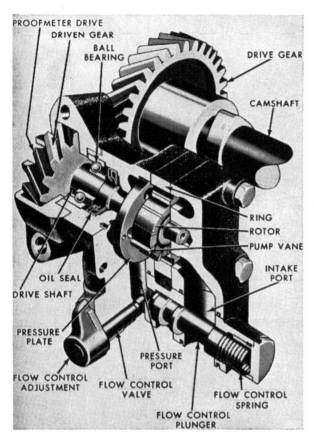

PROOFMETER DRIVE
DRIVEN GEAR
BALL
BEARING
DRIVE GEAR
CAMSHAFT
RING
ROTOR
PUMP VANE
INTAKE
PORT
OIL SEAL
DRIVE SHAFT
PRESSURE
PLATE
PRESSURE
PORT
FLOW CONTROL
ADJUSTMENT
FLOW CONTROL
VALVE
FLOW CONTROL
SPRING
FLOW CONTROL
PLUNGER

Fig. FO273 — Cutaway view of Vickers vane type hydraulic pump used on early production tractors. Later models use a piston type pump shown in Fig. FO275.

Refer to cross-sectional view of early production piston type pump in Fig. FO275 and to exploded view of late production piston pump in Fig. FO276. Early production pump is equipped with one double-row ball bearing on driveshaft; late production pump has one tapered roller bearing (12 & 13) and one needle bearing (10). To disassemble the piston type pump, proceed as follows:

Unbolt drive housing (8) from pump body (19) and withdraw drive unit (items 1 through 16) and "O" ring (17). Remove the six pump pistons (20) and six piston return springs (21).

To disassemble drive unit, remove wobble plate (16) and thrust bearing (15). Remove the Proofmeter drive unit (items 1 through 4) from drive housing and remove slotted head cap screw (5) and washers (6 and 7) from rear end of drive shaft (14). Insert a long, thin punch in hole in rear end of drive shaft and drive the shaft out of housing. Remove the tapered roller bearing cone (late production) or double row ball bearing from drive shaft with suitable pullers. On early production pumps, use a punch to drive the oil seal (11) out of housing. On late production pumps, insert special bearing remover (Nuday tool No. N-881-A) in needle bearing as shown in Fig. FO278. CAUTION: Be careful not to drive tool in too far; tap on tool lightly until a "click" is heard. If tool is accidentally driven in too far, it can be released from

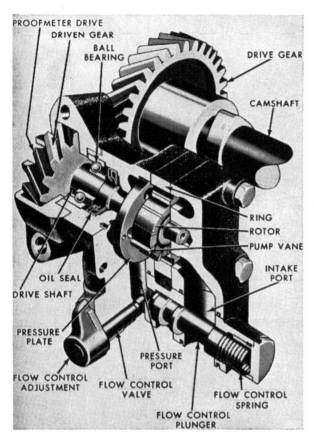

PRESSURE
HOLES

SEAL

Fig. FO274 — When installing the vane type pump shaft oil seal, use a tool that contacts only the outside diameter of seal.

327. OVERHAUL PISTON PUMP.
NOTE: When installing piston pump as replacement for vane pump, a new pump manifold, new engine oil pressure line and adapters must be installed. Be sure that piston pump clears oil line adapter elbow in cylinder block. Grind off adapter or rethread deeper into block if necessary.

Discard all "O" rings. If bore of cam ring is uneven or has a sharp ridge, or if rotor or any of its vanes are worn, install a new vane, rotor and cam ring kit. The shaft bushing is renewable separately from the cover. Assemble the shaft oil seal to body with the two pressure holes facing toward the drive gear as shown in Fig. FO274.

Assemble the cam ring with the cast-on arrow facing in the direction of pump rotation or so arrow on ring is pointing upward and is visible when pump is mounted on tractor. Refer to paragraph 325 for installation of pump to engine.

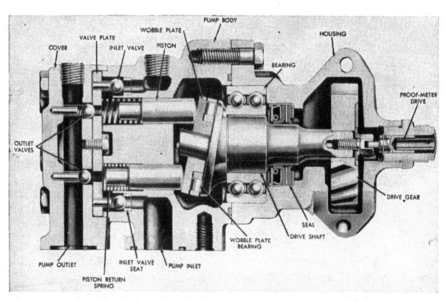

COVER
VALVE PLATE
INLET VALVE
WOBBLE PLATE
PISTON
PUMP BODY
HOUSING
BEARING
PROOF-METER DRIVE
OUTLET VALVES
DRIVE GEAR
SEAL
DRIVE SHAFT
WOBBLE PLATE BEARING
PUMP OUTLET
INLET VALVE SEAT
PUMP INLET
PISTON RETURN SPRING

Fig. FO275 — Sectional view of early type piston type pump. Refer to Fig. FO277 for bearing set-up used on later production piston pumps.

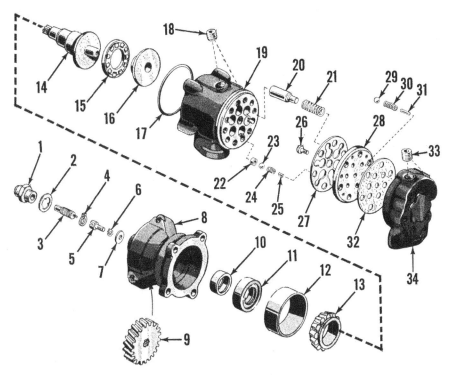

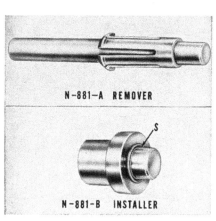

Fig. FO277 — Special tools required to remove needle bearing (10—Fig. FO276), oil seal (11) and bearing cup (12) from drive housing of late production piston type pump and to install new needle bearing. Refer to text.

Fig. FO276 — Exploded view of later production piston type pump using tapered roller and needle bearing on pump drive shaft. Body (19) is no longer fitted with pipe plugs (18), but is cast solid without openings.

1. Bushing	10. Needle bearing	18. Pipe plugs	26. Screw
2. Gasket	11. Oil seal	(no longer used)	27. Gasket
3. Drive shaft	12. Bearing cup	19. Pump body	28. Valve plate
(for Proofmeter)	13. Bearing cone &	20. Pistons (6)	29. Outlet valves (6)
4. Snap ring	roller	21. Return springs (6)	30. Valve springs (6)
5. Special screw	14. Pump drive shaft	22. Inlet valve seats (6)	31. Pins (6)
6. Lock washer	15. Thrust bearing	23. Inlet valves (6)	32. Gasket
7. Flat washer	16. Wobble plate	24. Valve springs (6)	33. Pipe plugs (2)
8. Drive housing	17. "O" ring	25. Pins (6)	34. Pump cover
9. Drive gear			

tapered bore in housing by tapping simultaneously outward and to the side to compress tool fingers on one side, thus providing clearance for finger hooks to pass by the tapered bore. Hammering on the tool as shown will drive out the needle bearing, oil seal and tapered roller bearing cup.

Unbolt cover (34) from body (19) and pry cover loose from gasket seal. Be careful not to lose any of the six inlet valves (steel balls) (23) or damage inlet valve springs (24). Extract the inlet valves from pump body and remove the valve springs from pins (25) that are pressed into the valve plate (28). Remove the screw (26) from center of valve plate and lay cover on work bench with valve plate down. Insert a punch through the oversize bolt hole in cover and tap valve plate loose from cover. Be careful not to lose the outlet valves (steel balls) (29). Remove the valve springs from pins (31) that are pressed into the cover. The inlet valve seats (22) can be removed from pump body with a special puller, Nuday tool No. NCA-600-G.

To reassemble pump, proceed as follows: Using special driver, Nuday tool No. NCA-600-EA or equivalent, drive the inlet valve seats into bores in pump body with sharp side of seats out.

Lay pump cover on bench, inner side up, and insert outlet valve springs (30) over the pins in cover. Place six steel balls (inlet and outlet steel valve balls are interchangeable) on the springs. Insert gasket (32) in cover so that all holes in gasket and cover are aligned. Place valve plate on gasket (See Fig. FO279) so that all holes in plate and gasket are aligned, drop two cap screws into the bolt holes to maintain alignment and install and tighten the retaining screw (26). Remove the aligning cap screws. Install inlet valve springs (24) on pins in valve plate with small end of springs toward the plate. Place pump body on bench, front end up, and drop six steel balls onto the inlet valve seats. Stick the gasket (27) to valve plate with grease so that all bolt holes are aligned and the inlet valve springs are in the slotted holes

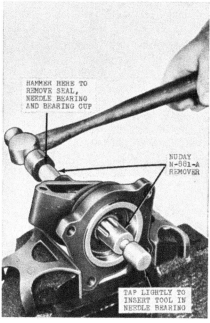

Fig. FO278 — Using special tool to remove needle bearing, oil seal and tapered roller bearing cup from late production piston type pump. Refer to Fig. FO277 and to text for procedure.

in gasket. Invert the cover assembly and carefully lower it over the pump body, being careful that the inlet valve springs enter the valve bores and that the pump to manifold mounting surfaces are aligned. Install the cover retaining cap screws and alternately tighten them to a torque of 40-50 Ft.-Lbs.

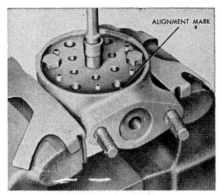

Fig. FO279—View showing alignment mark on valve plate and gasket.

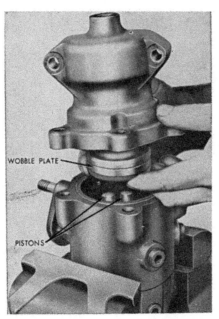

Fig. FO280 — Method in installing wobble plate on pistons so it will stay in position.

Turn pump body over and place it on bench with cover down. Insert the piston return springs in the piston bores. Lubricate the pistons with hydraulic oil, insert them in their bores with the crowned (rounded) ends up. Actuate all the pistons to be sure they are working freely.

On early production pumps, press double row ball bearing onto shaft by pressing on inner race only, until inner race is tight against shoulder on shaft. Press new oil seal into housing with spring loaded lip of seal towards rear (Proofmeter drive end) of housing. Lubricate lips of seal with Lubriplate, place drive gear in housing with flat side towards seal and insert shaft through oil seal and gear. Install and tighten the screw (5), lockwasher (6) and flat washer (7).

On late production pumps, place new needle roller bearing (10) on special installer (Nuday tool No. N-881-B) or equivalent tool, with lettered end of bearing cage towards driving shoulder (S—Fig. FO277) of tool. Drive or press bearing into place so that lettered end of bearing cage is slightly below flush with shoulder in housing; shoulder on tool N-881-B provides correct depth. Install oil seal (11—Fig. FO276) with spring loaded lip towards rear (Proofmeter drive end) of housing and with front face of seal $\frac{1}{16}$-inch below flush with shoulder in housing. CAUTION: Installing either the needle bearing or oil seal too far into housing will seriously damage either part. Drive bearing cup (12) into housing until tight against shoulder and drive bearing cone and roller (13) onto shaft (14) until tight against shoulder. Pack needle bearing and lubricate lips of oil seal with Lubriplate. Place drive

gear (9) in housing with flat side of gear towards needle bearing and insert shaft through bearings, seal and gear. Install and tighten screw (5), lockwasher (6) and flat washer (7).

Install assembled drive unit to pump body using a new "O" ring (17) as shown in Fig. FO280. Tighten the retaining cap screws to a torque of 40-50 Ft.-Lbs. Install pump on tractor and check gear backlash as outlined in paragraph 325 before installing Proofmeter drive unit. Lubricate Proofmeter drive shaft (3—Fig. FO276) and adapter (1) with Lubriplate, and install on pump drive housing using a new gasket (2).

328. PUMP MANIFOLD. On all models except Row Crop and Offset type tractors, the procedure for removal of manifold consists in draining the hydraulic reservoir and loosening (but not removing) the pump-to-engine bolts. Remove manifold-to-pump nuts and manifold to transmission cap screws then lift off the manifold. Discard the "O"rings.

On Row Crop and Offset type tractors, the procedure is the same except that it will be necessary to either remove the right side rail or to remove necessary bolts from same that will permit bottom of rail to be swung outwards enough to clear the pump-to-manifold flange.

NOTE: At time of introduction of diesel tractor, steel plate between transmission and engine was reduced in thickness on all models. If necessary to renew piston pump manifold, use Part No. NCA-933-J on 600, 700, 800 and 900 tractors and 601, 701, 801 and 901 prior to Serial Number 14257. On all models after 1958-62 production tractor Serial Number 14257, use manifold Part No. 310878.

329. RELIEF VALVE. Pump relief valve (94—Fig. FO269) is located in the bottom of the center (differential) housing in a vertically bored hole which is accessible after sump cover plate (97) has been removed from the bottom of the center housing. Bottom end of valve is internally threaded for a ¼-inch coarse thread screw or stud to facilitate removal. One removal method would be to use a piece of pipe of suitable inside diameter, long screw, large plain washers and a nut to pull the valve from the housing. Valve is adjusted at the factory (and then sealed with solder) to unseat at 2000-2200 psi gage reading and is serviced only as an assembly.

Unseating pressure can be checked as outlined in paragraph 304.

330. HYDRAULIC TUBES. Intake and outlet (pressure) tubes extend through the transmission to the hydraulic sump in the center housing. Tubes are a drive fit in transmission and are sealed at each end by "O" rings located in counterbores at both ends of transmission. Tubes can be driven out from rear and driven in from front after transmission is removed from tractor, by using piloted drifts such as Nuday tool Nos. NCA-945-A and NCA-945-B.

The oil pressure and return tubes (86 and 87—Fig. FO269) in center housing can be renewed after removing the lift cover assembly as outlined in paragraph 314, draining the hydraulic fluid and removing the plate (97) and pressure port plug (92B) from bottom of center housing. Drive the tubes downward until they are free of top flange of center housing, bend tubes over and extract them from bottom of center housing. Drive new tubes into place from top until they are flush with "O" ring counterbore in top face of center housing. Coat threads of plug (92B) with white lead and reinstall tightly. Reinstall plate (97) using new gasket (96) and "O" rings. Reinstall lift cover assembly as outlined in paragraph 314.

REMOTE CONTROL VALVES

Three different types of remote control valves are available, and are installed on the hydraulic lift cover after removing the accessory plate (11—Fig. FO263).

311877 Single or Double Acting Remote Control Valve

The 311877 remote control valve can be used with either single or double acting remote cylinders. When using a single acting cylinder, the rear port of the control valve is plugged with a special hex-head plug and sealing washer. A screw type valve is provided in the valve body to allow "float" action of a remote cylinder; closing the screw type valve will prevent oil from returning to the sump from remote cylinder until control valve is acutated. Due to shuttle type check valve operation, implements controlled by remote cylinder cannot be lowered unless engine is running, or unless the front port is pressurized and the screw type valve is opened.

331. ADJUSTMENT. For proper valve operation, the "kick-off" pressure (pressure in remote hydraulic lines at which remote control valve will automatically return to neutral position) should be adjusted to 1725-1800 psi. To adjust the "kick-off" pressure setting, proceed as follows:

Connect a hydraulic line and shut-off valve to either remote control valve pressure port. Direct the return flow back to sump. Remove plug (92A—Fig. FO269) from lower right hand side of center housing and insert a 0-3000 psi. gage in the plug opening. Start engine and operate hydraulic system until hydraulic fluid is at normal operating temperature. With shut-off valve in test hydraulic line open, move control valve handle to pressurized that port. (Check to see that fluid is running from line back into sump.) With engine running at

1200 RPM, gradually close the shut-off valve in hydraulic test line and observe pressure gage reading. The control valve should automatically return to neutral position when the pressure reaches 1725-1800 psi. If control valve returns to neutral position before the pressure reaches 1725 psi, tighten the adjusting screw (21—Fig. FO281). If pressure reaches 1800 psi before the control valve returns to neutral, loosen the adjusting screw (21). Note: If "kick-off" pressure cannot be brought up to 1725 psi and shut-off valve in test line is completely closed, a worn hydraulic pump or malfunctioning system relief valve should be suspected.

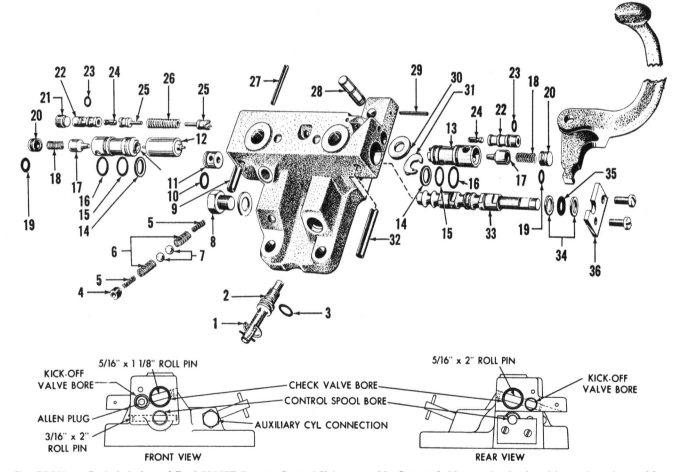

Fig. FO281 — Exploded view of Ford 311877 Remote Control Valve assembly. Detents hold control valve in raising or lowering position until cylinder reaches end of stroke, then valve automatically "kicks off" to neutral position.

1. Snap ring	7. Detent balls	13. Check valve bushing	19. "O" rings	25. Spacers	31. Spring clip
2. Float valv	8. Accessory plug	14. Back-up rings	20. Plugs	26. Spring	32. Roll pin
3. "O" ring	9. Roll pin	15. "O" rings	21. Adjusting screw	27. Roll pin	33. Control valve
4. Plug	10. "O" ring	16. "O" rings	22. Valve bushings	28. Lever pin	34. Flat washers
5. Detent springs	11. Plug	17. Check valves	23. "O" rings	29. Roll pin	35. "O" ring
6. Detent springs	12. Shuttle plunger	18. Springs	24. "Kick-off" valves	30. Flat washer	36. Retainer plate

135

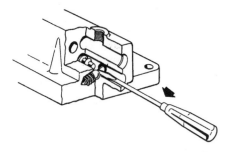

Fig. FO282 — View showing one method of holding lower detent ball and springs in place while inserting control valve spool in bore.

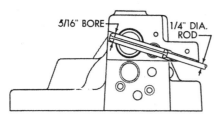

Fig. FO283 — Removing check valve retaining roll pins.

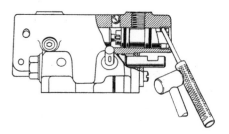

Fig. FO284 — Removing check valve assemblies.

332. R&R AND OVERHAUL. To remove the remote control valve, first disconnect any remote hoses, place the lift system selector lever in draft control position and move the touch control lever to bottom of quadrant. Stand on lift arms to exhaust all oil from lift cylinder; then, unbolt and remove remote control valve from lift cover. Renew the sealing "O" rings between valve and lift cover when reinstalling valve and coat the valve and lift cover mating surfaces with grease.

333. CONTROL LEVER. To remove the remote control valve lever, move lever forward to raise position and drive the ⅛-inch roll pin (29—Fig. FO281) out far enough to front of valve body so that the pin (28) can be removed. Then, remove lever from valve body.

To reinstall lever, remove spring clip (31) and flat washer (30) from pin (28). Position lever on valve body with tang on lever engaged in notch in control valve spool (33). Insert pin (28) through lever and valve body with grooved end of pin down, align groove in center of pin with roll pin (29) and drive roll pin in flush with valve body. Place flat washer (30) over bottom end of pin (28), then insert retaining clip, cupped side up, in groove in bottom end of pin.

334. CONTROL VALVE SPOOL. To renew the "O" ring seal (35) on control valve spool, remove lever as outlined in paragraph 333; then, remove plate (36), one flat washer (34) and the "O" ring. Lubricate new "O" ring with Lubriplate and install by reversing removal procedure.

To remove the control valve spool, remove plug (4) springs (5 & 6, one each) and one detent ball (7). With lever and plate (36) removed as in preceding paragraph, withdraw the valve spool, catching the second detent ball (7) and remove second set

of detent springs (5 & 6) as valve is removed. If necessary to remove plug (11) or renew "O" ring (10), drive the ³⁄₁₆-inch roll pin (27) out of valve body. Then, insert a clean wood dowel through valve spool bore and push plug out to front side of valve.

To reinstall valve spool proceed as follows: Be sure valve bore is clean, then insert one set of springs (5 & 6, with small spring inside large spring) in detent hole and drop one detent ball (7) on top of springs. Depress the detent ball and springs as shown in Fig. FO282 while inserting valve spool in bore. Note: If plug (11) has not been removed the detent ball and springs can be depressed with a small punch through the plug (4) hole. Insert the second detent ball (7) and second set of springs (5 & 6) on top of valve, then install plug (4) tightly. Install "O" ring on plug (11), lubricate "O" ring with Lubriplate and insert plug, "O" ring end first, into front end of valve bore. Align cross hole in plug with hole in valve body, then drive the roll pin (27) through valve body and plug. Insert one flat washer (34) lubricate "O" ring (35) and install over valve spool, place second washer (34) over "O" ring and install retaining plate (36) and the control valve lever.

335. "KICK-OFF" VALVES. Remove control valve handle as in paragraph 333. Unscrew plug (21), counting the turns (or part of turn) required to bring plug out flush with valve body. Then, remove the plug and, using a ¼-inch rod, drive the ⁵⁄₁₆-inch roll pin (32) past the "kick-off" valve bore. The "kick-off" valve bushings (22), valves (24), spacers (25) and spring (26) can then be removed from valve body.

Wash all parts in solvent and renew any that are damaged. To reinstall proceed as follows: Center the spring (26) and two spacers (25) in the valve bore. Renew "O" rings (23) on the valve bushings (22) and, using grease, stick the valves, round end

first, into the bushings. Lubricate the "O" rings with Lubriplate and insert the rear valve and bushing just far enough into the bore so that the roll pin (32) can be driven back into place. Then, insert the front valve and bushing and the plug (21). CAUTION: Pushing or driving the "kick-off" valve bushings into the ends of the valve bore any farther than necessary will cut the sealing "O" rings (23) and result in an oil leak. Turn the plug (21) in same number of turns (or part of turn) past flush with valve body as it was before being removed. Note: If position of plug was not noted before removal, screw the plug in just flush with valve body. In any case, the "kick-off" pressure must be readjusted as outlined in paragraph 331.

336. CHECK VALVES. After removing the "kick-off" valves as outlined in paragraph 335, drive the roll pins (9 & 32) out of the valve body. The check valve bushings (13) with valves (17), springs (18) and plugs (20) can then be removed as shown in Fig. FO284. Remove check valves from the bushings as shown in Fig. FO285.

Be sure the steel check valve seats (See Fig. FO285) are in correct location in valve bushing. The seats are not renewable separately from the bushings; if seats are not located properly or are damaged, renew the bushings. Inspect shuttle plunger (12 —Fig. FO281) and renew if scored or if plunger tips are broken or otherwise damaged.

To reassemble, proceed as follows: Install heavy "O" ring (16) in groove at outer end of bushing and install thin "O" ring (15) in groove at inner end of bushing. Install back-up ring (14) in groove on inner side of thin "O" ring (15). Insert check valve (17) and spring (18) in bushing. Install new "O" ring (19) on plug (20), lubricate the "O" ring and push plug into bushing with flat side out. Lubricate the "O" rings on outside of bushing. In-

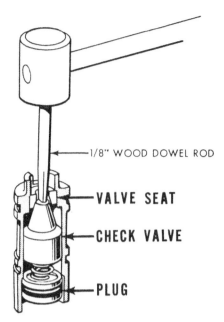

—1/8" WOOD DOWEL ROD

VALVE SEAT

CHECK VALVE

PLUG

Fig. FO285 — Removing check valves from bushings.

sert both bushing assemblies in valve body with plunger (12) centered between them and align the roll pin holes in the bushings with holes in valve body. Install the roll pins; be sure that pin (9) does not protrude into "kick-off" valve bore and install pin (32) in side of valve body as indicated by dotted line in Fig. FO281 only far enough to retain the check valve bushing.

Reinstall "kick-off" valves as outlined in paragraph 335 and proceed with reassembly of valve as outlined in previous paragraphs. After valve is reinstalled on tractor, readjust the "kick-off" pressure as outlined in paragraph 331.

337. "FLOAT" VALVE. To renew the "O" ring seal (3) on float valve (2), or to renew the float valve, proceed as follows: Turn float valve in, remove snap ring (1); then, turn the valve completely out of valve body. The float valve or sealing "O" ring can then be renewed.

338. **TROUBLE SHOOTING.** Before suspecting malfunction of remote control valve first be sure that the tractor hydraulic system is functioning properly and that pump (and relief valve) is capable of delivering a minimum of 2000 psi. If tractor hydraulic system is ok, refer to following paragraphs:

WILL NOT OPERATE CYLINDER. If control valve will not move, service control valve and detent mechanism as outlined in paragraph 334. If

valve will move to each detent position, and tractor hydraulic system is ok, service "kick-off" valves as outlined in paragraph 335 and service check valves as outlined in paragraph 336.

PREMATURE "KICK-OFF". If control valve returns to neutral position before remote cylinder has reached end of stroke, first adjust the "kick-off" pressure as outlined in paragraph 331. Be sure that detent plug (4—Fig. FO281) is tight. Note: Valve will "kick-off", if properly adjusted, whenever pressure in remote hydraulic line reaches 1725-1800 psi. If valve seems ok, check for obstruction in hydraulic line, use of too small I.D. hose or overloading of remote cylinder.

WILL NOT "KICK-OFF". Check or adjust "kick-off" pressure as outlined in paragraph 331. Be sure that tension on detent springs is not excessive; loosen plug (4) if this seems likely.

"KICK-OFF" CYCLING. If control valve "kicks-off" past the neutral detent position, check "kick-off" pressure as outlined in paragraph 331. Remove any "home made" attachments from control lever; extra weight on lever will carry lever past the neutral detent. Check action of detent balls (7) and springs (5 & 6). Be sure that tractor engine is not being operated at too high RPM.

REMOTE CYLINDER WILL NOT HOLD POSITION. Be sure the "float" valve screw (2—Fig. FO281) is tight. Service the check valves as outlined in paragraph 336 and install new "O" ring kit.

290207 Single Spool Remote Control Valve and 290124 Double Spool Remote Control Valve

Cross-sectional views of the 290207 Single Spool and 290124 Double Spool Remote Control Valves are shown in Figs. FO287 and FO288. Either valve can be used with single or double acting remote cylinders. On Single Spool Valve, open (turn out) switch valve (See Fig. FO288) to operate a single acting cylinder from "LIFT" port of valve. On Double Spool Valve, back the Allen head screw (See Fig. FO287) out two or three turns to operate a single acting remote cylinder from the "No. 1 LIFT" port. Single Spool and Double Spool Remote Control Valves do not have detent positions: valve levers must be held in raising or lowering position. Spools are spring loaded and will return to neutral position when control valve lever is released. If valve is held open after remote cylinder reaches end of stroke, a relief valve (poppet valve) within the remote control valve body will open at 1900-2000 psi.

339. **ADJUSTMENT.** Relief valve setting within the remote control valve should be adjusted to 1900-2000 psi as follows: Connect a 3000 psi pressure gage to any valve port as shown in Fig. FO286. Bring hydraulic fluid to normal operating temperature. When the port in which the gage is connected is pressurized, the gage reading should be 1900-2000 psi. Pressure setting can be increased by addition of shims (See Fig. FO287 or FO288). Addition of one shim will increase pressure setting approximately 150 psi.

Fig. FO286 — View showing pressure gage installed in port of Double Spool Control Valve.

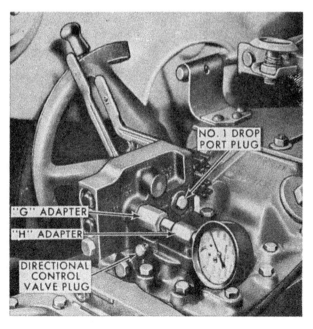

NO. 1 DROP PORT PLUG

"G" ADAPTER

"H" ADAPTER

DIRECTIONAL CONTROL VALVE PLUG

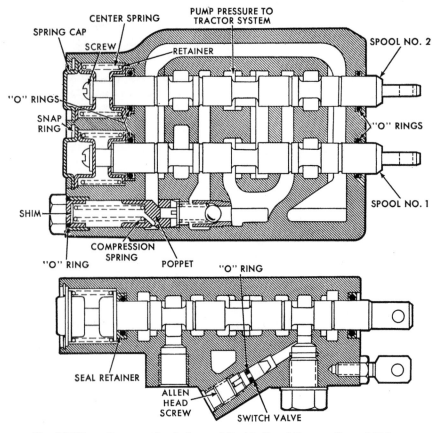

Fig. FO287 — Cross-sectional views of Double Spool Remote Control Valve.

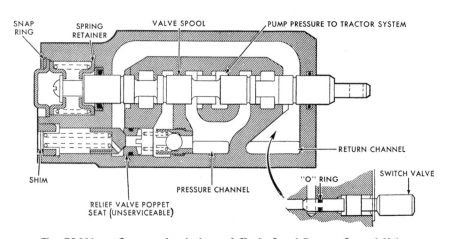

Fig. FO288 — Cross-sectional views of Single Spool Remote Control Valve.

340. R&R AND OVERHAUL. Refer to paragraph 332 for removal and reinstalling procedure. Refer to following paragraphs for overhaul procedure.

341. CONTROL VALVE SPOOLS. Place a wooden block between the valve spools or between valve spool and work bench and carefully drive out the spool handle roll pins. Remove the handles. Remove the spool retaining snap rings and centering spring caps. Gently tap the spools on handle end and remove through front of valve body. On Double Spool Valve, tag each valve spool as each is of a selected fit. Withdraw the metal spool seal ("O" ring) retainers from front end of valve body with remover (Nuday tool No. 7600-E) or equivalent. Remove and discard the "O" rings from front and rear of valve body. Clean the spools and attaching parts thoroughly.

If the control valve spools or spool bores are seriously damaged or scored, the entire remote control valve must be renewed as spools are not serviced. Renew centering springs, retainers or washers if cracked or distorted. Centering spring free length should be 1.103; renew spring if free length varies materially from this dimension.

Reinstall control valves as follows: Lubricate the control valve spools and insert from front end of valve body. Note: Be sure that the valves are reinstalled in correct bores in Double Spool Valve. Push the valve spools rearward so that rear end of spool is aligned with rear "O" ring groove; this will prevent "O" ring from being dropped into valve passage. Lubricate the rear "O" rings, and using valve spool as "backstop", insert "O" rings in grooves of valve body. Push the valve spools through the "O" rings so that spools extend about ¾-inch out rear end of valve body. Lubricate the front "O" rings and install the "O" rings and metal retainers with seal driver (Nuday tool No. N-651) or equivalent. Reinstall the centering springs, retainer screws, washers, centering spring caps and valve handles.

342. RELIEF VALVE. Remove the relief valve cap, shims, valve spring and valve poppet from control valve body. Free length of valve spring should be 1.731; renew spring if free length varies materially from this dimension. Renew relief valve poppet if scored or grooved. Poppet seat is not serviced. Readjust relief valve setting as outlined in paragraph 339 after valve is reinstalled on tractor.

343. SINGLE SPOOL SWITCH VALVE. Remove control valve handle, handle adjustment link and switch valve retaining lock from valve body. Turn the one-piece switch valve entirely out with a ⅜-inch Allen wrench. Remove and discard sealing "O" ring. Renew valve if excessively pitted or distorted. Install and lubricate new "O" ring, then reinstall valve in valve body.

344. DOUBLE SPOOL SWITCH VALVE. Drive the Allen head screw retaining roll pin inward into body bore and remove the pin. Remove the Allen head screw. Remove switch valve poppet by tilting and tapping valve body. Renew the valve poppet if pitted or distorted. Renew the sealing "O" ring, lubricate and reinstall valve, Allen head plug and plug retaining roll pin.

HYDRAULIC POWER PACKAGE

Heavy Duty Industrial

Heavy duty industrial tractors are optionally fitted with a hydraulic power package, consisting of a crankshaft driven gear type pump, separate reservoir and control valve with attaching lines suitable for operating loader, backhoe or other mounted industrial equipment. The gear type pump is driven by an adapter on the front of the engine crankshaft. The five gallon oil reservoir is mounted above the pump in the grille housing. The stack type valve is mounted on a bracket on the right side of the tractor and normally consists of two spool sections. Service procedures are as follows:

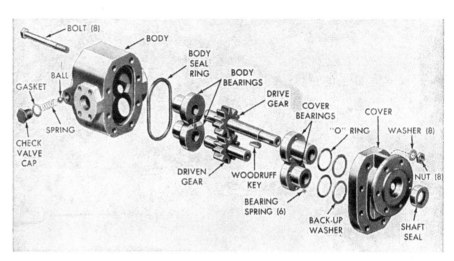

Fig. FO291 — Exploded view of 17 gpm gear type pump used on the hydraulic power package. Body bearings and gears are serviced only as matched units. All other parts are available separately.

RESERVOIR AND LINES

345. All steel hydraulic lines are fitted with straight tube threads with "O" ring seals or straight tube threads with 37-degree flared fittings. To remove the reservoir, unbolt and remove the grille unit as outlined in paragraph 151, unbolt and remove the headlights and withdraw the reservoir from the rear of the grille casting. To drain the system, place a drain pan under the filter element and remove the element. Note: A drain plug is provided at the bottom of the reservoir, but a suitable trough to direct the drained oil out of the grille enclosure is difficult to fashion, and the filter element will still need to be removed to drain the lines below the level of the reservoir drain plug. Fig. FO290 shows a cross-sectional view of the filter assembly showing location of the by-pass valve which operates if the filter should become plugged.

When installing tubing, do not tighten any connection until all connections are started. Always renew the sealing "O" ring when reinstalling the lines.

HYDRAULIC PUMP

346. **REMOVE AND REINSTALL.** To remove the pump, first remove the grille door and unbolt and remove the two split type flanges on the pump pressure line. Remove the pressure tube from the pump, disconnect the inlet hose clamp and remove the inlet hose. Remove the four nuts and lockwashers retaining the pump mounting flange to the front support and withdraw pump, mounting flange and drive shaft from the front of the tractor.

When reinstalling the pump, position the pump on the mounting flange studs with the drive shaft extending through the hole in the front support. Install the drive adapter on the pump drive shaft splines; then, install and tighten the four nuts and lockwashers to the front support. With the pump in place, align the adapter with the mounting holes in the crankshaft hub and install the securing bolts. Always renew the "O" ring seal on the pump pressure line when reinstalling the pump.

347. **OVERHAUL.** After removal of the pump as outlined in paragraph 346, thoroughly clean the exterior of the pump with a suitable solvent.

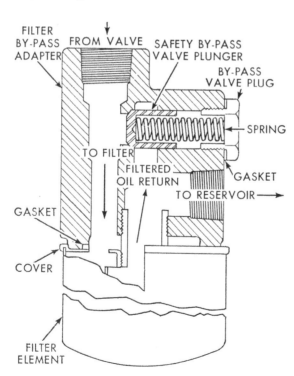

Fig. FO290 — Cross sectional view of power package oil filter showing location and action of bypass valve which acts if the cartridge becomes plugged.

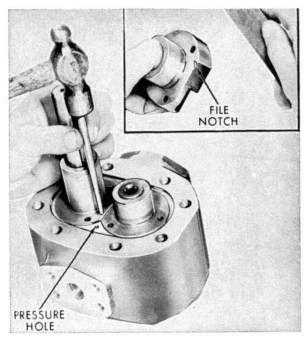

Fig. FO292 — Staking the cover bearings when reassembling the pump. See text for details.

shaft bores and applying pulling pressure on bearings. Remove and examine check valve and seat. Check valve operates to limit pressure on shaft seals to 10 psi.

Gears and body bearings are available in matched sets if renewal is indicated. All other parts are available individually. Wash all parts thoroughly in a suitable solvent and examine for wear or other damage.

Parts should be reassembled absolutely clean and dry. Install body bearings separately with relief groove to pressure port side of body, Bearings should have 0.0005 clearance in body to provide good hydraulic seal. Install drive and driven gears in body. Renew gears or body if more than 0.007 radial clearance exists between gear teeth and body. Drive gear is installed in left body bearing, when facing pressure port side of body as shown in Fig. FO292. Install drive gear cover bearing in body and file a notch ⅛ inch deep in driven gear bearing as shown in inset. Notch should be placed on pressure port side, on outer surface of bearing. Clean and dry driven gear bearing and intall in body. Seat both bearings

NOTE: The same standards of care and cleanliness necessary in the handling of any hydraulic or diesel components should be observed in the disassembly and service of the hydraulic pump.

Remove the inlet flange and gasket, mounting plate and universal drive shaft from the pump and place the pump in a soft jawed vise with the drive shaft up. Refer to Fig. FO291. Remove the Woodruff key from the drive shaft and remove the nuts and washers from the pump cover bolts. Examine the exposed end of the pump drive shaft and remove any visible nicks or burrs; then, carefully remove the pump cover from the pump. If cover bearings are removed with the cover, do not let them drop out on a hard or sharp surface. The cover "O" rings and pump shaft seal can be renewed at this time. When renewing the seal, stake the outer edge of the seal to the pump cover in three places to firmly anchor the seal.

Remove the six cover bearing springs and cover bearings. Note: If the cover bearings cannot be readily withdrawn, tap the pump body lightly with a plastic hammer while pulling on bearings. Never use a hard faced hammer.

Remove the drive and driven gears from the body bore and remove the body bearings by placing thumbs in

Fig. FO293 — Control valve outer end plate showing method of plugging pump return for operation of additional valves in series. Port closed by plug (B) then becomes pressure line for added valves.

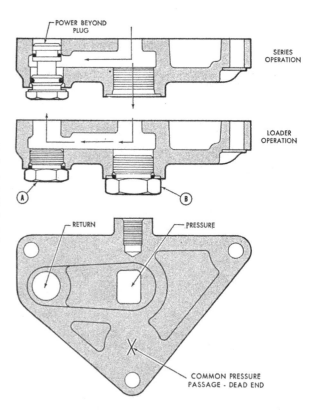

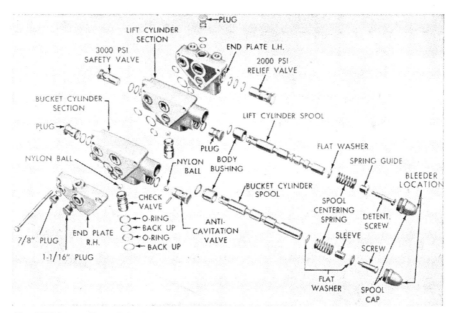

Fig. FO294 — Exploded view of Cessna control valve uesd on the hydraulic power package. Anti-cavitation valve, the two relief valves and two valve body plugs will all interchange. For proper valve operation they must be installed in the locations shown.

firmly, then use a center punch and stake drive gear bearing into notch previously filed in driven gear bearing as shown. Tap just hard enough to displace metal in the filed notch. Coat the cover bearing "O" rings and back-up rings with Lubriplate and install in cover. Coat seal lip and drive shaft with hydraulic oil and reinstall cover. Torque the retaining bolts to 40-42 Ft.-Lbs. after first making sure that the bearing springs are all properly in place and the cover can be pressed down in position flush with the body with a reasonable effort.

Reinstall the mounting plate and inlet flange. Reinstall the woodruff key and slide the universal coupling on the pump shaft. Note: Do not drive the coupling on the pump shaft. If difficulty is encountered remove any burrs on shaft, key and coupling until it can be pushed into place. Install coupling on pump shaft so that there is $\frac{5}{16}$-inch clearance between rear yoke shoulder and hub of drive coupling (Refer to Fig. FO297) and lock front yoke in place with set screw. Yoke must be positioned as outlined to maintain clearance between end of pump shaft and U-joint pressure gun fitting (See Fig. FO297).

Fill pump with oil and turn by hand before reinstalling on tractor.

CONTROL VALVE

The Cessna sectioned control valve is normally serviced with a lift and bucket control spool section in addition to the two valve end plates. The left hand end plate contains the pressure and return lines as well as the cartridge type system relief valve. The right hand end plate is fitted with two closure plugs in the passage leading from the valve pressure port to the return port as shown in Fig. FO293. When plug (A) is removed and fitted with a "power beyond" plug as shown in upper view, the center port becomes a pressure port leading to a second auxiliary valve for backhoe and other operations. Provisions are made for auxiliary valve oil return in a manifold block adjacent to the system oil filter. The lift cylinder valve section is fitted with a 3,000 psi cylinder safety valve and float position detent. Both valves contain a nylon lift check valve and the bucket control valve is fitted with an anti-cavitation valve to maintain a solid bucket cylinder during the dumping operation.

348. SYSTEM PRESSURE CHECK AND TROUBLE SHOOTING. Refer to Fig. FO294 for the proper order of

parts installation. It is possible to install the 2,000 psi system relief valve, the 3,000 psi cylinder safety valve, the anti-cavitation valve and the two valve section end plugs in the wrong ports in the valve bodies. Care must be taken to see that they are correctly installed in their correct locations. The pressure relief and safety valves are stamped on the hex head with the factory pressure setting. Individual parts of the relief valves are not serviced and the relief valve cartridges should not be disassembled. If found to be defective, they must be renewed as a unit. Renewal of the relief valves and plug "O" rings can be accomplished without removal of the valve unit from the tractor if the proper standards of care and cleanliness are observed. To check the system pressure, attach a 3,000 psi pressure gage into a tee in the lift cylinder cross-over line and with the system at operating temperature, increase the engine speed to 1500 rpm. Raise the lift cylinders to the limit of their stroke and note the gage reading while holding the valve lever in the raising position. Gage reading should be 1,950-2,050 psi. If a new relief valve does not correct the condition and the relief pressure drops quickly, overhaul the pump as outlined in

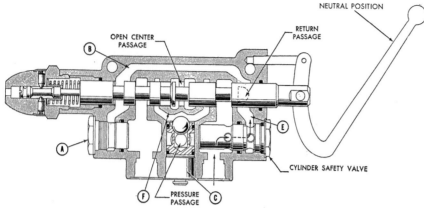

Fig. FO295 — Cross sectional view of lift valve section showing oil flow and action of cylinder safety valve. Oil enters the valve through central pressure passage in lift check valve plug (C) and returns to sump through end passage (B) and (E). Cylinder safety valve acts in the event of shock loads to return lift cylinder oil to return lines at (E). Open center passage is to provide a float position when the valve is detented in the lowering position.

unit and are not available individually. Remove valve spools from the cap ends of the valve bodies after removing the caps and levers. Mark the valve spools and bodies to make sure that spools are installed in the correct body. Always renew "O" rings and seals when the valve is reassembled.

paragraph 347. In conditions of erratic action or settling of the units the valve check balls and seat and/or the anti-cavitation valve should be suspected as well as the valve sealing "O" rings and the cylinder packing.

349. **R&R AND OVERHAUL.** Removal procedure of the control valve assembly is evident. The lift check valve seats are retained in the valve bodies by the lower through-bolt. Valve spools and bodies are matched

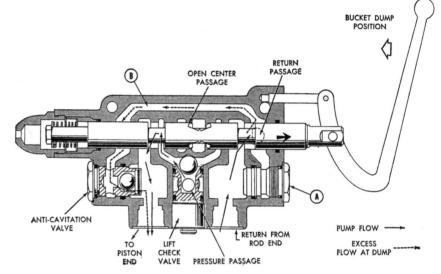

Fig. FO296 — Cross sectional view of bucket control section of valve showing operation of anti-cavitation valve. When lever is moved to the dump position as shown, oil from the pump lifts check valve from seat and flows to piston end of dump cylinder. If weight of bucket forces oil from rod end of cylinder faster than pump delivery, part of dump oil flows through anti-cavitation valve to keep piston end of cylinder full.

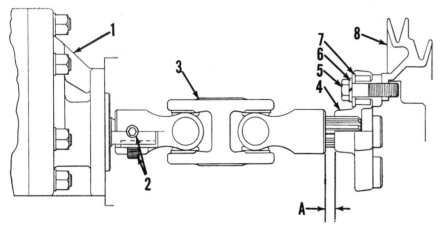

1. Hydraulic pump
2. Set screws
3. Drive shaft
4. Adapter hub
5. Cap screws
6. Lock washers
7. Flat washers
8. Crankshaft pulley

Fig. FO297 — Drawing showing Heavy Duty Industrial hydraulic power package pump shaft. When installing pump and/or shaft, it is important to maintain 5/16-inch clearance at "A" to prevent excessive end loading of engine crankshaft, and subsequent damage to center main bearing liner and crankshaft thrust surfaces.

350. WIRING DIAGRAMS.

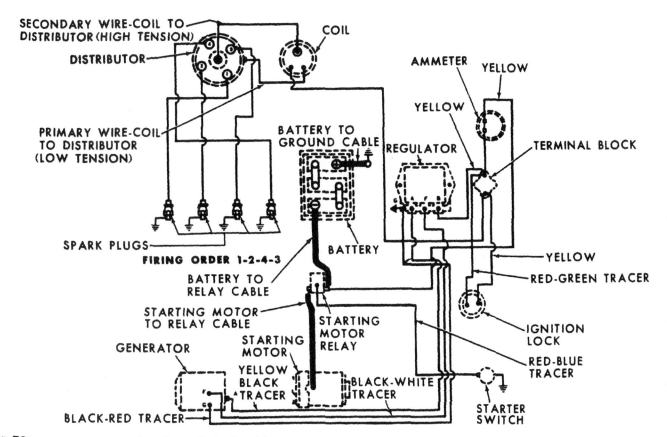

Fig. FO298 – Electrical system wiring diagram for 6-volt models.

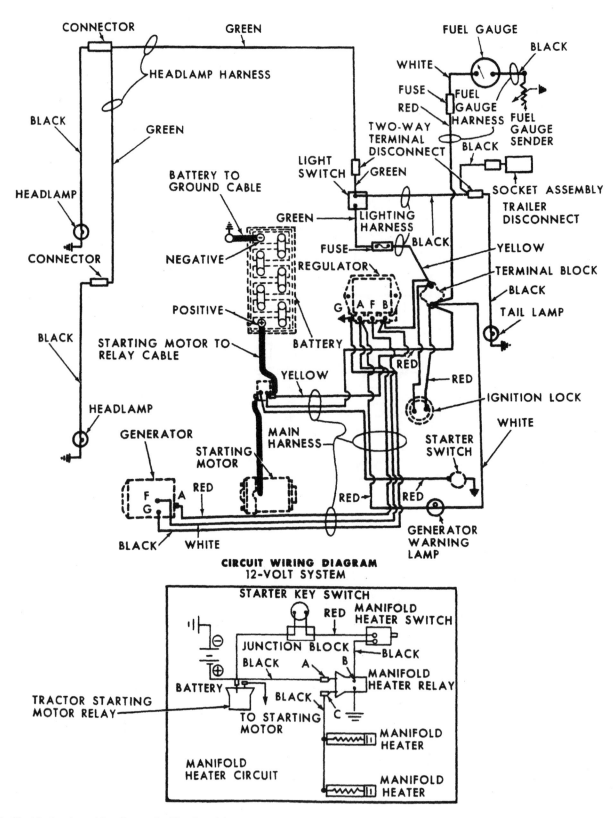

Fig. FO299 – Electrical system wiring diagram for 12-volt models.